Electrical Design for Building Construction

Electrical Design for Building Construction

John E. Traister

John E. Traister Associates

McGraw-Hill Book Company

New York St. Louis San Francisco Auckland Düsseldorf
Johannesburg Kuala Lumpur London Mexico Montreal New Delhi
Panama Paris São Paulo Singapore Sydney Tokyo Toronto

Library of Congress Cataloging in Publication Data

Traister, John E
    Electrical design for building construction.

    Includes index.
    1. Electric lighting—Drawings. 2. Electric wiring—Diagrams.
    3. Buildings—Electric equipment—Drawings. I. Title.
TK4169.T7   621.319'24   76-14971
ISBN 0-07-065127-2

1234567890   HDHD   785432109876

The editors for this book were Tyler G. Hicks and Ruth L. Weine, the
designer was Richard A. Roth, and the production supervisor was
Frank P. Bellantoni. It was set in VIP Memphis Medium by
Progressive Typographers.

Printed and bound by Halliday Lithograph Corporation

Contents

Preface

The analysis of existing drawings prepared by professional engineers has long been a major part of the "serious" engineering student's method of study. Unfortunately, the young electrical engineering student who hopes to become an engineer/designer in a consulting engineering firm has no wealth of material to aid him. Printed working drawings of electrical systems in buildings are rare. The basic purpose of this book is to remedy this situation by giving the budding professional a means of seeing and analyzing actual working drawings of electrical systems used in residential, commercial, institutional, and industrial structures.

We will not concern ourselves with too much theory as hundreds of previous books have taken ample care of this for us. Our main concern is the application of practical electrical designs for light and power that must be dealt with by the professional engineers in consulting firms.

Practically all of the examples given are in working-drawing form with written explanations to clearly illustrate at a glance exactly what is taking place and also for what purpose. No further data are necessary to understand the examples except the written specifications found in Appendix A of this book.

Our examples are taken from these projects:

Residence
J. C. Bowman, ARA, Architect
Harrisonburg, Va.

Tenant Residence
Baughan & Baukhages, AIA, Architects
Luray, Va.

The First National Bank—Branch
Baughan & Baukhages, AIA, Architects
Luray, Va.

Laundry Building
G. Lewis Craig, Architect
Waynesboro, Va.

The First National Bank—Main Office
Baughan & Baukhages, AIA, Architects
Luray, Va.

Church
Dwight E. Miller, AIA, Architect
Harrisonburg, Va.

Department Store
Baughan & Baukhages, AIA, Architects
Luray, Va.

Nursing Home
Baughan & Baukhages, AIA, Architects
Luray, Va.

A deep and grateful bow must be made in the direction of the draftspersons, designers, and engineers who prepared these drawings. I am especially indebted to Boyd G. Headley, III, who prepared many of the illustrations for this book, and to Ruby R. Updike, whose encouragement and many hours of typing helped make this project much easier than it otherwise would have been.

No matter how many books a person has read on the subject and no matter how many sets of working drawings he has seen and analyzed, his progress can only be measured by what he actually designs and what designs have been constructed.

If this book can take a few of the stumbling blocks out of the student's path and light the way somewhat, my purpose will have been fulfilled.

John E. Traister

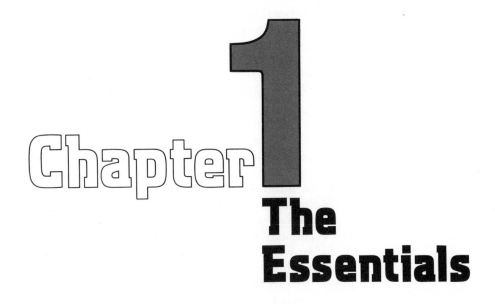

Chapter 1
The Essentials

OBJECTIVES

The fundamental objective of this chapter is to give the reader an overall picture of the electrical construction industry and the electrical designer's relationship to it. Familiarity with these relationships is considered necessary to give designers a proper background for approaching their work more intelligently.

This chapter will also introduce fundamental design procedure, electrical symbols, and other basic essentials necessary to the electrical design profession.

THE HISTORICAL RELATION OF THE ELECTRICAL DESIGNER TO THE BUILDING CONSTRUCTION INDUSTRY

The earliest recorded use of commercial electricity was the installation of an arc lamp in the Dungeness lighthouse in England in 1862. This early type of light, while not entirely steady or

1

free from smoke, was able to produce great amounts of very bright light—for the time—by drawing electric arcs or flames between two carbon electrodes. However, it was not until seventeen years later, in 1879, that the first successful carbon-filament incandescent lamp was invented by Thomas A. Edison. He also developed the first efficient electric generator to supply current for his lamps in the same year. The experiments were continued, and in 1882 Edison developed the first central-station electric generating plant in New York City.

Prior to Edison's generating plant, naturally there was no inclusion of electrical systems in building construction. However, this new generating plant brought on a public demand for the use of electric lighting and power in existing buildings, as well as for new construction.

These first electrical wiring installations were usually laid out by workers employed and trained by the power companies, and the majority of these installations were "designed" by the mechanics on the job—often as the work progressed. Building contractors then began hiring mechanics of their own to install the electrical wiring systems, but because of the special skills and knowledge required, these same builders soon began leaving the wiring installations to mechanics who began to specialize in this work as electrical contractors. In fact, most of the building contractors were more than happy to subcontract this work since the electrical installations of that time pertained to only a small percentage of the overall contract.

Owing to the potential hazards caused by the improper handling and installation of electrical systems, certain rules in the selection of materials, quality of workmanship, and precautions for safety had to be followed in order to minimize these hazards. The National Electrical Code (NEC) was prepared in 1897 to standardize and simplify these rules and also to provide some reliable guide for electrical construction. It is recommended that the reader obtain the latest edition of the NEC for reference in all design projects.

As the electrical construction continued to become a more important part of the general building construction, the architects began to prepare layouts of the desired electrical systems on the architectural drawings. This layout usually indicated the lighting outlets, base "plugs," and light switches by means of certain symbols. A line was sometimes drawn from a lighting outlet to a wall switch to indicate how the various lamps were to be controlled, but this was usually the extent of the electrical design. The details of wiring, number of circuits, etc., were still left to the mechanics installing the system. As the electrical systems became more extensive and complex, the electrical contractors began hiring draftsmen to prepare working drawings to supplement the sketchy outlet layout on the architectural drawings, to provide a basis for preparing estimates, and to give instructions to electricians in the field who were installing the systems. These early electrical draftsmen were probably experienced architectural or mechanical draftsmen who were trained in the field of electricity by the electrical contractors.

From that point on, electrical construction continued to become a more important part of general building construction, and soon the architects began to prepare more extensive layouts of the electrical systems until finally separate drawings were included along with the architectural drawings. As the volume of such layout work increased and the electrical systems became still more extensive and complex, a greater engineering knowledge of power and illumination requirements became necessary. Persons with the proper knowledge and training began to devote their time exclusively to designing and laying out electrical installations as consulting engineers, selling their services to the architects. These consulting engineers conveyed their designs by means of working drawings which used symbols, lines, notations, etc. Thus, the electrical designer has become a very important cog in the wheel of electrical construction.

THE CONSTRUCTION OF A BUILDING

In nearly all instances, when an owner decides to have a building constructed, an architect is hired to prepare the complete working drawings and specifications for the building. The drawings usually include (1) a plot plan indicating the location of the building on the property; (2) elevations of all exterior faces of the building; (3) floor plans showing the walls and partitions for each floor or level; (4) sufficient vertical cross sections to indicate clearly the various floor levels and details of the foundation, walls, floors, ceilings, and roof construction; and (5) large-scale renderings showing such details of construction as may be required.

For jobs of any consequence, the architect usually includes drawings and specifications covering the design of plumbing, heating, ventilating, air conditioning, and electrical work. A brief description of such drawings follows.

Mechanical Drawings

The mechanical drawings cover the complete design and layout of the plumbing, piping, heating, ventilating, air conditioning, and related mechanical construction. They also include floor-plan layouts, cross sections of the building, and necessary detailed drawings. The control wiring for mechanical equipment is usually indicated on the mechanical drawings, not on the electrical drawings as one might imagine.

Electrical Drawings

The electrical drawings generally cover the complete electrical design of the electrical wiring for lighting, power, signals and communications, special electrical systems, and related electrical equipment. These drawings should also include a plot plan or site plan showing the location of the building on the property and the interconnecting electrical systems; floor plans showing the location of power outlets, lighting fixtures, panelboards, etc.; power-riser diagrams; a symbol list; schematic diagrams; and larger-scale details where necessary.

The architect will often represent the owner in soliciting quotations from general contractors and advise the owner as to the proper award to make. The architect will also usually represent the owner during construction of the building, inspecting the work to ascertain that it is being performed in accordance with the requirements of the drawings and specifications.

The relations of the consulting engineer parallel those of the architect, inasmuch as the consulting engineer often represents the architect and owner in soliciting from the contractors quotations which pertain to the engineer's line of work. The engineer also inspects his or her portion of the work to assure the architect that this portion is carried out according to the working drawings and specifications. The engineer approves shop drawings (material submittals), checks and approves progress payments, and performs similar duties—all pertaining only to the engineer's phase of the construction.

Architects sometimes require the engineer to prepare an approximate estimate of the electrical work to aid them in determining the probable cost of the building prior to the actual request of formal quotations. This is especially true on government projects. The book *Rapid Electrical Estimating & Pricing*, written by C. Kenneth Kolstad and published by McGraw-Hill Book Company, is a handy reference for this type of electrical estimating.

Quality of Designs

The consulting engineering firms should strive to produce designs and working drawings which are no less than excellent in quality. They should not only be correct as far as wiring methods, feeder sizes, etc., are concerned, but they should also be presented in such a manner as to be easily interpreted by the electrical contractor, the electrical estimators, and the electricians on the job. The electrical drawings combined with the written specifications should leave no doubt as to exactly what is required of the electrical contractor for the proper installation of the electrical system on the project.

ELECTRICAL WIRING SYMBOLS

The purpose of an electrical drawing is to show the complete design and layout of the electrical systems for lighting, power, signal and communication systems, special raceways, and related electrical equipment. In preparing such drawings, the electrical layout is shown through the use of lines, symbols, and notation which should indicate, beyond any question or any doubt, exactly what is required.

Many engineers, designers, and draftsmen use symbols adapted by the United States of America Standards Institute (USASI). However, no definite standard schedule of symbols is always used in its entirety. Consulting engineering firms quite frequently modify these standard symbols to meet their own needs. Therefore, in order to identify the symbols properly, the engineer provides, on one of the drawings or in the written specifications, a list of symbols with a descriptive note for each—clearly indicating the part of the wiring system which it represents.

Figure 1-1 shows a list of electrical symbols which are currently recommended by USASI, while Fig. 1-2 shows another list of symbols which was prepared by the Consulting Engineers Council/U.S. and the Construction Specifications Institute, Inc.

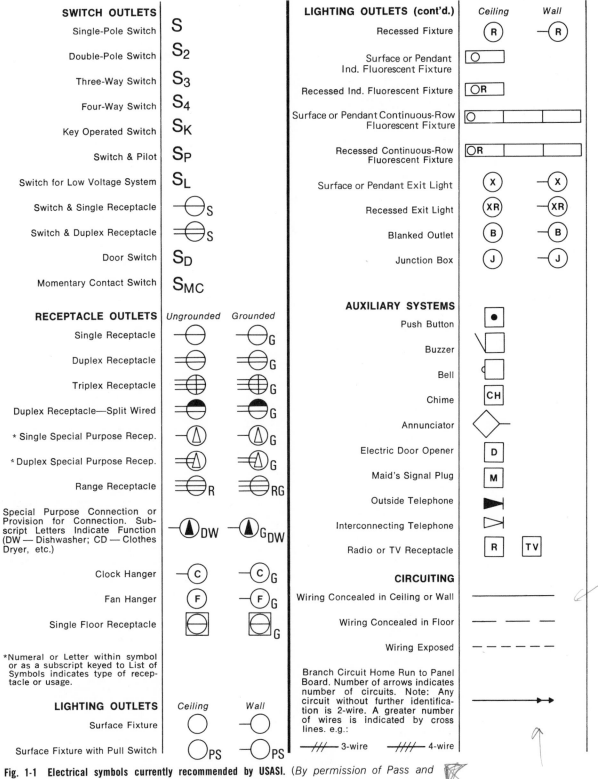

Fig. 1-1 Electrical symbols currently recommended by USASI. (*By permission of Pass and Seymour, Inc.*)

Electrical Reference Symbols

Electrical Abbreviations

(Apply only when adjacent to an electrical symbol)

Central switch panel	CSP
Dimmer control panel	DCP
Dusttight	DT
Emergency switch panel	ESP
Empty	MT
Explosionproof	EP
Grounded	G
Night light	NL
Pull chain	PC
Raintight	RT
Recessed	R
Transfer	XFER
Transformer	XFRMR
Vaportight	VT
Watertight	WT
Weatherproof	WP

Switch Outlets

Single-pole switch	S
Double-pole switch	S$_2$
Three-way switch	S$_3$
Four-way switch	S$_4$
Key-operated switch	S$_K$
Switch and fusestat holder	S$_{FH}$
Switch and pilot lamp	S$_P$
Fan switch	S$_F$
Switch for low-voltage switching system	S$_L$
Master switch for low-voltage switching system	S$_{LM}$
Switch and single receptacle	S
Switch and duplex receptacle	S
Door switch	S$_D$
Time switch	S$_T$
Momentary contact switch	S$_{MC}$
Ceiling pull switch	S
"Hand-off-auto" control switch	HOA
Multi-speed control switch	M
Push button	

Receptacle Outlets

Where weatherproof, explosionproof, or other specific types of devices are to be required, use the upper-case subscript letters. For example, weatherproof single or duplex receptacles would have the uppercase WP subscript letters noted alongside of the symbol. All outlets should be grounded.

Single-receptacle outlet	
Duplex-receptacle outlet	
Triplex-receptacle outlet	
Quadruplex-receptacle outlet	
Duplex-receptacle outlet— split wired	
Triplex-receptacle outlet— split wired	
250-V receptacle single-phase— Use subscript letter to indicate function (DW, dishwasher; RA-range; DC, clothes dryer) or numeral (with explanation in symbol schedule)	
250-V receptacle—three-phase	
Clock receptacle	C
Fan receptacle	F
Floor single-receptacle outlet	
Floor duplex-receptacle outlet	
Floor special-purpose outlet*	
Floor telephone outlet—public	
Floor telephone outlet—private	

Example of the use of several floor outlet symbols to identify a 2, 3, or more gang floor outlet:

Underfloor duct and junction box for triple- double-, or single-duct system as indicated by the number of parallel lines.

Example of the use of various symbols to identify location of different types of outlets or connections underfloor duct or cellular floor systems:

Cellular floor header duct

Circuiting

Wiring exposed (not in conduit)	—E—
Wiring concealed in ceiling or wall	
Wiring concealed in floor	
Wiring existing*	
Wiring turned up	
Wiring turned down	
Branch-circuit home run to panelboard.	1 2

*Use numeral keyed to explanation in drawing list of symbols to indicate usage.

*Note: Use heavyweight line to identify service and feeders. Indicate empty conduit by notation CO (conduit only).

Bus Ducts and Wireways

Trolley duct‡	T	T
Busway (service, feeder, or plug-in)‡	B	B
Cable-trough ladder or channel‡	C	C
Wireway‡	W	W

Panelboards, Switchboards, and Related Equipment

Flush-mounted panelboard and cabinet‡	
Surface-mounted panelboard and cabinet‡	
Switchboard, power-control center, unit substations (should be drawn to scale)‡	
Flush-mounted terminal cabinet (In small-scale drawings the TC may be indicated alongside the symbol)	TC
Surface-mounted terminal cabinet (In small-scale drawings the TC may be indicated alongside the symbol)	TC
Pull box (Identify in relation to wiring-system section and size)	
Motor or other power controller (may be a starter or contractor)‡	
Externally operated disconnection switch‡	
Combination controller and disconnection means‡	

Power Equipment

Electric motor (hp as indicated)	¼
Power transformer	
Pothead (cable termination)	
Circuit Element, e.g., circuit breaker	CB
Circuit breaker	
Fusible element	
Single-throw knife switch	
Double-throw knife switch	
Ground	
Battery	
Contractor	C
Photoelectric cell	PE
Voltage cycles, phase	Ex: 480/60/3
Relay	R
Equipment connection (as noted)	

Remote-Control Stations for Motors or Other Equipment

Push-button station	PB
Float switch, mechanical	F
Limit Switch, Mechanical	L
Pneumatic switch, Mechanical	P
Electric eye, beam source	
Electric eye, relay	
Temperature-control relay connection (3 denotes quantity.)	R 3
Solenoid-control valve connection	S
Pressure-switch connection	P
Aquastat connection	A
Vacuum-switch connection	V
Gas-solenoid-valve connection	G
Flow-switch connection	F
Timer-connection	T
Limit-switch connection	L

Lighting

	Ceiling	Wall
	Type	Switch
Surface or pendant incandescent fixture PC = pull chain		PC Circuit
Surface or pendant exit light		
Blanked outlet	B	B
Junction box	J	J
Recessed incandescent fixtures		
Surface or pendant individual fluorescent fixture		
Surface or pendant continuous-row fluorescent fixture (letter indicating controlling switch)		
	O	A

Fixture No.
Wattage

Symbol not needed at each fixture

*Bare-lamp fluorescent strip

†Note: Any circuit without further identification indicates two-wire circuit. For a greater number of wires, indicate with cross lines, e.g.:

3 wires: 4 wires, etc.

Neutral wire may be shown longer. Unless indicated otherwise, the wire size of the circuit is the minimum size required by the specification. Identify different functions of wiring system, e.g., signaling system by notation or other means.

‡Identify by notation or schedule

Electric Distribution or Lighting System, Aerial

Pole	
Street or parking-lot light and bracket	
Transformer	
Primary circuit*	
Secondary circuit**	
Down guy	
Head guy	
Sidewalk guy	
Service weather head	

Electric Distribution or Lighting System, Underground

Manhole	M
Handhole	H
Transformer manhole or Vault	TM
Transformer pad	TP
Underground direct burial cable (Indicate type, size and number of conductors by notation or schedule.)	
Underground duct line (Indicate type, size, and number of ducts by cross-section identification of each run according to notation or schedule. Indicate type, size, and number of conductors by notation or schedule.)	
Street-light standard fed from Underground circuit	

Signaling System Outlets

Institutional, Commercial, and Industrial Occupancies

I Nurse call-system devices (any type)

Basic symbol

Examples of individual item identification (not a part of Standard):

Nurses' annunciator (add a number after it as 24 to indicate number of lamps	①24
Call station, single-cord, pilot light	②
Call station, double-cord, microphone speaker	③
Corridor dome light, 1 lamp	④
Transformer	⑤
Any other item on same system— use numbers as required.	⑥

II Paging-system devices (any type)

Basic symbol

Examples of individual item identification (not a part of Standard):

Keyboard	①
Flush annunciator	②
2-Face annunciator	③
Any other item on same system— use numbers as required.	④

III Fire-alarm-system devices (any type) including smoke- and sprinkler-alarm devices

Basic symbol

Examples of individual item identification; (not a part of Standard)

Control panel	①
station	②
10" Gong	③
Presignal chime	④
Any other item on same system— use numbers as required.	⑤

IV Staff-register-system devices (any type)

Basic symbol

Examples of individual item identification:; (not a part of Standard):

Phone-operators' register	①
Entrance register, flush	②
Staff-room register	③
Transformer	④
Any other item on same system— use numbers as required.	⑤

V Electric clock-system devices (any type)
Basic symbol

*In the case of continuous-row bare-lamp fluorescent strip above an area-wide diffusing means, show each fixture run, using the standard symbol; indicate area of diffusing means and type by light shading and/or drawing notation.

†Identify by notation or schedule

Examples of individual item identification, (not a part of Standard):

Master clock	①
12" Secondary, flush	②
12" Double dial, wall-mounted	③
18" Skeleton dial	④
Any other item on same system— use numbers as required.	⑤

VI Public telephone-system devices

Basic symbol

Examples of individual item identification, (not a part of Standard):

Switchboard	1
Desk phone	2
Any other item on same system— use numbers required.	3

VII Private telephone system Devices (any type)

Basic symbol

Examples of individual item identification, (not a part of Standard):

Switchboard	1
Wall phone	2
Any other item on same system— use numbers as required.	3

VIII Watchman-system devices (any type)

Basic symbol

Examples of individual item identification. (not a part of Standard):

Central station	①
Key station	②
Any other item on same system— use numbers as required.	③

IX Sound system

Basic symbol

Examples of individual item identification, (not a part of Standard):

Amplifier	①
Microphone	②
Interior speaker	③
Exterior speaker	④
Any other item on same system— use numbers as required.	⑤

X Other signal-system devices

Basic symbol

Examples of individual item identification (not a part of Standard)

Buzzer	①
Bell	②
Push button	③
Annunciator	④
Any other item on same system— use numbers as required.	⑤

Residential Occupancies

Signaling system symbols for use in identifying standardized residential-type signal system items on residential drawings where a descriptive symbol list is not included on the drawing. When other signal system items are to be identified, use the above basic symbols for such items together with a descriptive symbol list.

Push button	
Buzzer	
Bell	
Combination bell-buzzer	
Chime	CH
Annunciator	
Electric door opener	D
Maid's signal plug	M
Interconnection box	
Bell-ringing transformer	BT
Outside telephone	
Interconnecting telephone	
Television outlet	TV

Fig. 1-2 (Facing page) A list of electrical symbols recommended by the Consulting Engineers Council/U.S. and the Construction Specifications Institute, Inc.

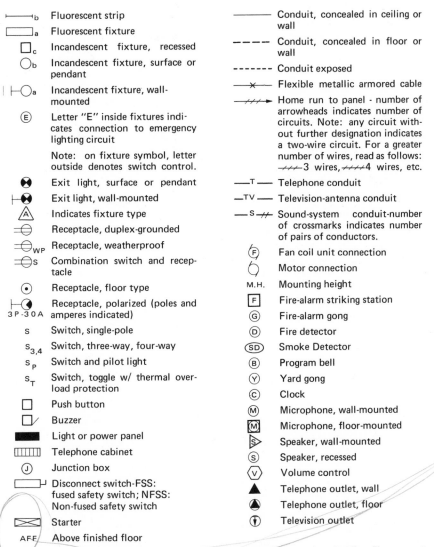

⊢———▪b	Fluorescent strip
▭a	Fluorescent fixture
▫c	Incandescent fixture, recessed
◯b	Incandescent fixture, surface or pendant
⊢◯a	Incandescent fixture, wall-mounted
Ⓔ	Letter "E" inside fixtures indicates connection to emergency lighting circuit
	Note: on fixture symbol, letter outside denotes switch control.
⊗	Exit light, surface or pendant
⊢⊗	Exit light, wall-mounted
△	Indicates fixture type
⊖	Receptacle, duplex-grounded
⊖wp	Receptacle, weatherproof
⊖s	Combination switch and receptacle
⊙	Receptacle, floor type
⊢◖ 3P-30A	Receptacle, polarized (poles and amperes indicated)
s	Switch, single-pole
s₃,₄	Switch, three-way, four-way
s_P	Switch and pilot light
s_T	Switch, toggle w/ thermal overload protection
▫	Push button
▫╱	Buzzer
▬	Light or power panel
⊞⊞⊞	Telephone cabinet
Ⓙ	Junction box
▭⌐	Disconnect switch-FSS: fused safety switch; NFSS: Non-fused safety switch
⊠	Starter
A.F.F.	Above finished floor

———	Conduit, concealed in ceiling or wall
————	Conduit, concealed in floor or wall
-------	Conduit exposed
——×——	Flexible metallic armored cable
——⫻/→	Home run to panel - number of arrowheads indicates number of circuits. Note: any circuit without further designation indicates a two-wire circuit. For a greater number of wires, read as follows: ⫻—3 wires, ⫻⫻—4 wires, etc.
——T——	Telephone conduit
——TV——	Television-antenna conduit
——S—⫽—	Sound-system conduit-number of crossmarks indicates number of pairs of conductors.
Ⓕ	Fan coil unit connection
◯	Motor connection
M.H.	Mounting height
Ⓕ	Fire-alarm striking station
Ⓖ	Fire-alarm gong
Ⓓ	Fire detector
ⓈⒹ	Smoke Detector
Ⓑ	Program bell
Ⓨ	Yard gong
Ⓒ	Clock
Ⓜ	Microphone, wall-mounted
Ⓜ	Microphone, floor-mounted
▷s	Speaker, wall-mounted
Ⓢ	Speaker, recessed
Ⓥ	Volume control
▲	Telephone outlet, wall
◉	Telephone outlet, floor
⊕	Television outlet

Fig. 1-3 A modified list of electrical symbols used by a consulting engineering firm. These are standard symbols and may not all appear on any one project drawing; however, wherever the symbol appears on a project drawing, the item shall be provided and installed.

Figure 1-3 shows still another list of electrical reference symbols which have been modified for use by one consulting engineering firm.

It is evident from the preceding symbols that many have the same basic form, but, because of some slight difference, their meaning changes. For example, the outlet symbols in Fig. 1-4 each have the same basic form—a circle—but the addition of a line or an abbreviation gives each an individual meaning. A good procedure to follow in learning symbols is to first learn the basic form and then apply the variations for obtaining different meanings.

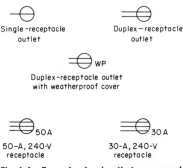

Single-receptacle outlet   Duplex-receptacle outlet

Duplex-receptacle outlet with weatherproof cover

50A   30A

50-A, 240-V receptacle   30-A, 240-V receptacle

Fig. 1-4 Example showing that many symbols have the same basic form but are slightly changed to give each an individual meaning.

The electrical symbols described in the following paragraphs represent a system of electrical notation whose compactness and clarity may be of assistance to electrical engineers, designers, and draftsmen.

The system should be used in place of standard symbols only if there seems to be a decided advantage in doing so.

Some of the symbols are abbreviated idioms, like "WP" for weatherproof and "AFF" for above finished floor. Other symbols are simplified pictographs, like ▷─◯─◁ for a double floodlight fixture or ⊟ for an infrared electric heater with two lamps.

In some cases that are combinations of idioms and pictographs, as in ⬛F₃₀ₐ for fusible safety switch, ⬛N₃₀ₐ for nonfusible safety switch, and ⬛DT₆₀ₐ for double-throw safety switch, where the pictograph of a switch enclosure is combined with the abbreviated idioms of F (fusible), N (nonfusible), and DT (double-throw), respectively. The numerals indicate the bus bar capacity in amperes.

This list came about as a result of much discussion with consulting engineers, electrical designers, electrical draftsmen, electrical estimators, electricians, and others who are required to interpret electrical drawings. It is felt that this list represents a good set of symbols in that they are:

1. Easy to draw
2. Easily interpreted by workers
3. Sufficient for most applications

Lighting Outlets

◯ Ceiling outlet with surface-mounted incandescent lighting fixture

◯⊣ Wall outlet with surface-mounted incandescent lighting fixture

Practical use of these symbols is shown in Fig. 1-5.

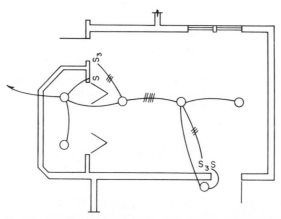

Fig. 1-5 Practical application of surface-mounted incandescent fixtures.

◎⊣ Wall outlet with recessed incandescent lighting fixture

◎ Ceiling outlet with recessed incandescent lighting fixture

Practical use of the symbols at the left is shown in Fig. 1-6. Sometimes these symbols are modified in order to indicate the physical shape of a particular incandescent fixture. The

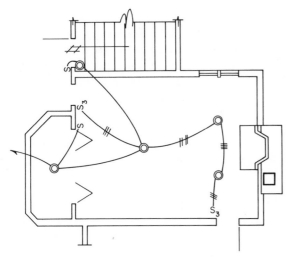

Fig. 1-6 Practical application of recessed incandescent lighting fixtures.

lighting fixture in Fig. 1-7 consists of four 6-in. cubes. This type of lighting fixture may be indicated on the electrical floor plan as shown in Fig. 1-8.

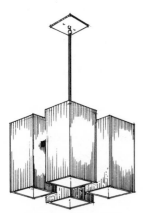

Fig. 1-7 Pictorial drawing of a lighting fixture consisting of four 6-in. cubes, one-cube lighting fixture on the drawings.

A lighting fixture consisting of one cube may be indicated as shown in Fig. 1-9. All should be drawn as close to scale as possible.

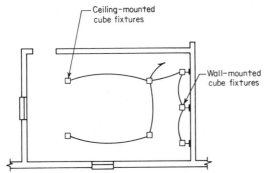

Fig. 1-8 Example showing how the lighting fixture in Fig. 1-7 may be indicated on working drawings.

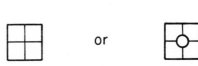

Fig. 1-9 Example showing how to indicate a one-cube lighting fixture on the drawings.

The type of mounting of all fixtures is usually indicated in the lighting fixture schedule shown on the drawings or in the written specifications. The fixture illustrated in Fig. 1-7 is obviously pendant-mounted and should be so indicated in an appropriate column in the lighting fixture schedule, since the floor-plan view in Fig. 1-8 does not indicate this fact.

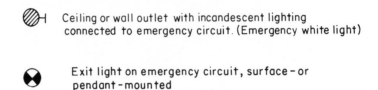

Ceiling or wall outlet with incandescent lighting connected to emergency circuit. (Emergency white light)

Exit light on emergency circuit, surface- or pendant-mounted

Exit light on emergency circuit, wall-mounted

Surface- or wall-mounted exit light with directional arrowheads

The mounting height of wall-mounted lighting fixtures is sometimes indicated in the symbol lists, especially where most are to be mounted at one height. For example, it might read ". . . wall outlet with incandescent fixture mounted 6 ft 6 in. above finished floor to center of outlet box unless otherwise indicated." If a few wall-mounted fixtures were to be mounted at 8 ft 0 in. above finished floor, they could be indicated as shown in Fig. 1-10, the letters AFF meaning "above finished floor."

+96 in. AFF   +96 in. AFF

Fig. 1-10 Method of indicating mounting height of a few wall-mounted fixtures.

Ground-mounted incandescent uplight

Post-mounted incandescent fixture

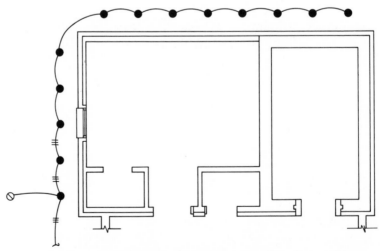

Fig. 1-11 Practical application of ground- and post-mounted incandescent fixtures.

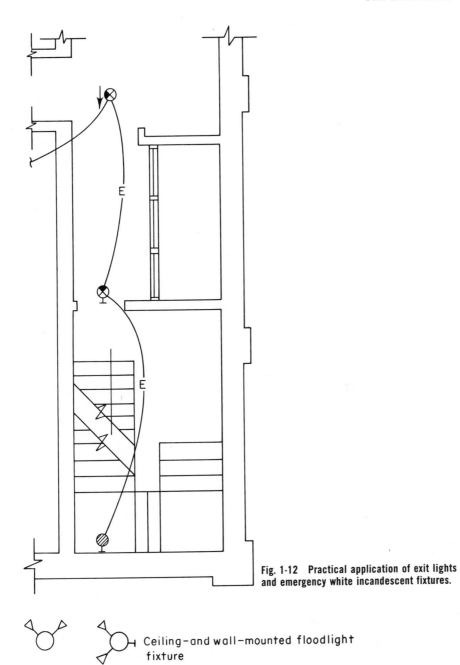

Fig. 1-12 Practical application of exit lights and emergency white incandescent fixtures.

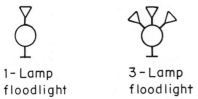

Ceiling-and wall-mounted floodlight fixture

If only one lamp (or more than two lamps) is required on the floodlight outlet, it can be shown as in Fig. 1-13 or Fig. 1-14.

1- Lamp
floodlight

3 - Lamp
floodlight

Fig. 1-13 Practical application showing floodlights on the drawings, either one lamp or more than two lamps.

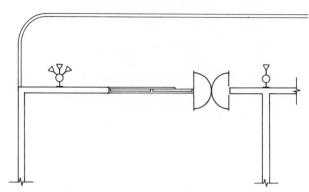

Fig. 1-14 Practical application of floodlights.

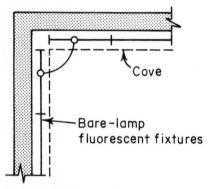

Fig. 1-15 Example showing how to indicate bare-tube fluorescent fixtures on the drawings.

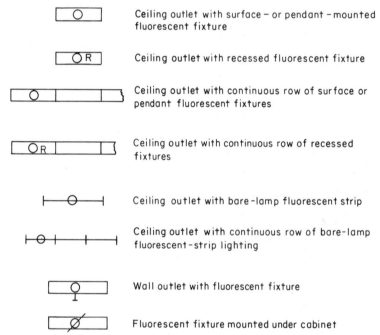

Ceiling outlet with surface – or pendant – mounted fluorescent fixture

Ceiling outlet with recessed fluorescent fixture

Ceiling outlet with continuous row of surface or pendant fluorescent fixtures

Ceiling outlet with continuous row of recessed fixtures

Ceiling outlet with bare-lamp fluorescent strip

Ceiling outlet with continuous row of bare-lamp fluorescent-strip lighting

Wall outlet with fluorescent fixture

Fluorescent fixture mounted under cabinet

Modification of the symbols for fluorescent lighting is common. For example, cove lighting with bare-lamp fluorescent strips may be indicated on the drawings as shown in Fig. 1-15.

The lighting layout illustrated in Fig. 1-16 shows practical applications of all fluorescent symbols covered in this writing.

Many electrical drawings do not differentiate between recessed, surface, or pendant-mounted fixtures on the floor plans. Rather, the mounting is indicated either in the lighting fixture schedule or in the written specifications. However, since a major variation in the type of outlet box, outlet supporting means, wiring system arrangement, and outlet connection, plus need of special items such as plaster rings or roughing-in cans, depends upon the way in which a fixture is mounted, the electrician should be able to know the type of mounting in a glance at the drawings. Therefore, the mounting of a lighting fixture should be indicated on the floor plans as well as in the lighting fixture schedule.

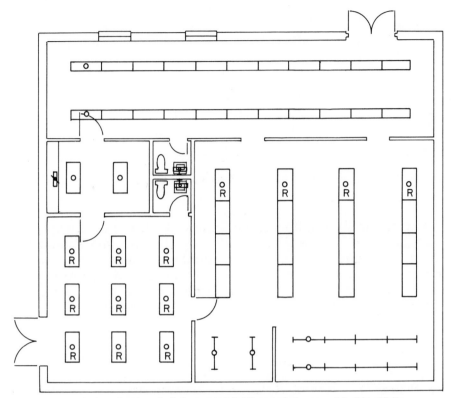

Fig. 1-16   Practical application of all fluorescent lighting symbols covered in this chapter.

Lighting fixtures are identified as to type of fixture by a numeral placed inside a triangle near each lighting fixture, as shown to the right and in Fig. 1-17.

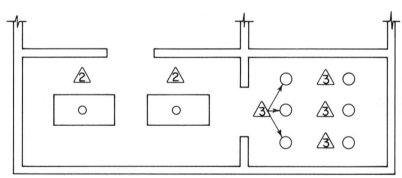

Fig. 1-17   Method of identifying type of lighting fixture.

The indicated fixture is shown in the symbol lists as follows:

⑤ Indicates type of lighting fixture — see schedule

If one type of fixture is used exclusively in one room or area, as shown in Fig. 1-18, the indicator need only appear once with the word "All" lettered at the bottom.

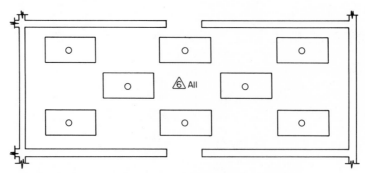

Fig. 1-18 Method of identifying several lighting fixtures of one type.

A complete description of the fixture identified by the symbol must be given in the lighting fixture schedule and should include the manufacturer, catalog number, number and type of lamps, voltage, finish, mounting, and any other necessary information needed for a proper installation of the fixtures. Figure 1-19 shows an example of a lighting fixture schedule.

LIGHTING FIXTURE SCHEDULE						
FIXT. Type	Manufacturer's Description	LAMPS No.	Type	Volts	Mounting	Remarks

Fig. 1-19 Example of a lighting fixture schedule.

Fluorescent fixtures should be drawn to approximate scale, showing physical size whenever practical.

Mercury vapor and other electric-discharge lighting fixtures are indicated on the drawings in the same way as incandescent fixtures. The type of lamp is indicated in the lighting fixture schedule.

Receptacles and Switches

Every day those who work in the electrical construction industry hear the word "standard." Yes, it would be the ideal situation if all electrical engineers could use one set of standard electrical symbols for all their projects. However, with the present symbols known as "standard" this is not practical.

For example, one consulting engineering firm in Washington, D.C., did a large amount of electrical designs for hospitals all over the eastern United States. On one of their jobs, there were over 100 duplex receptacles mounted horizontally in the backsplash of lavatory countertops, while there were over 300 conventional duplex receptacles.

If a draftsman had to letter a note at each of the receptacles located in the backsplash, it is easily seen that much time would be spent on the drawings. On the other hand, a simple symbol with a written explanation in the legend would tell exactly what work was to be done, and the draftsman could indicate the location of these outlets on the drawings in little, if any, extra time.

Our firm recently designed the electrical systems for a group of quick-service restaurants where several of the duplex receptacles were located in the ceiling. If we had used standard symbols, each location would have looked like the following:

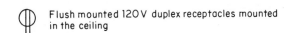 Flush mounted 120V duplex receptacles mounted in the ceiling

We saved much time by composing a new symbol for the ceiling-mounted receptacles and eliminated the descriptive note at each of the dozen or so receptacles. A note appeared only once in the symbol list as follows:

 Flush-mounted 120 V, duplex "twist-lock" receptacle mounted in ceiling with stainless steel plate

The cases are endless, and until a sufficient number of different symbols is available in the standard symbol list, it is highly impractical to use the standard throughout.

If a special outlet is shown on a drawing in only a few locations, then perhaps a descriptive note at each location is best; but if the "special outlet" appears in several locations, then an individual symbol can save much drawing time.

The following are switch and receptacle symbols used in our office on working drawings.

S Single-pole toggle switch mounted 50 in. AFF to center of box unless otherwise indicated.

S_2 Two-pole toggle switch mounted 50 in. AFF to center of box unless otherwise indicated.

S_3 Three-way switch mounted 50 in. AFF to center of box unless otherwise indicated.

S_4 Four-way switch mounted 50 in. AFF to center of box unless otherwise indicated.

S_L Low-voltage switch to relay mounted 50 in. AFF unless otherwise indicated.

S_D Flush-mounted door switch to control closet light.

S_P Switch with pilot light.

Figure 1-20 illustrates some practical applications of switch symbols used on working drawings. Notice that the single-pole switches are used to control lighting from one point, while the three- and four-way switches, used in combination, control a light or series of lights from two or more points.

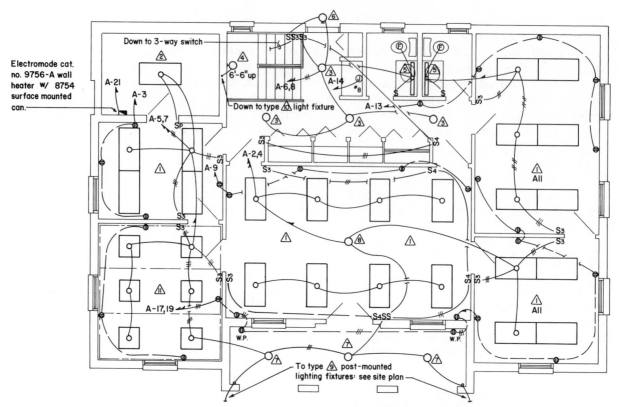

Fig. 1-20 Practical applications of switch symbols used on working drawings.

The two-pole switch is used to control a series of lights on two separate circuits with only one motion. The switch–pilot light combination is used where it is practical to notice if the item controlled by the switch is energized, such as a light in an attic or closet.

Door switches to control closet lights are quite common in residential construction. When the closet door is opened, the switch button is released and in turn energizes the circuit to the closet light. When the door is closed, the light is de-energized.

Receptacle symbols used in consulting engineering firms are numerous. Some firms have used over 50 different symbols for receptacles and power outlets. However, our firm has been able to work with only six different symbols to cover most of our needs. They are as follows:

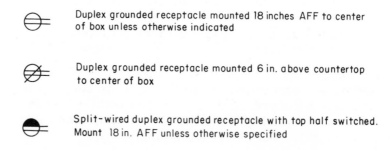

Duplex grounded receptacle mounted 18 inches AFF to center of box unless otherwise indicated

Duplex grounded receptacle mounted 6 in. above countertop to center of box

Split-wired duplex grounded receptacle with top half switched. Mount 18 in. AFF unless otherwise specified

3-pole, 3-wire, 240-V receptacle, amperage as indicated
30a

 Special outlet or connection, letter indicates type; see legend
a   at end of symbol list

Floor outlets are indicated on our drawings by a square with the appropriate symbol drawn inside:

Floor-mounted single receptacle, grounded

Floor-mounted duplex receptacle, grounded

J   Floor-mounted junction box

Figure 1-21 shows some practical applications of receptacle symbols as used on working drawings.

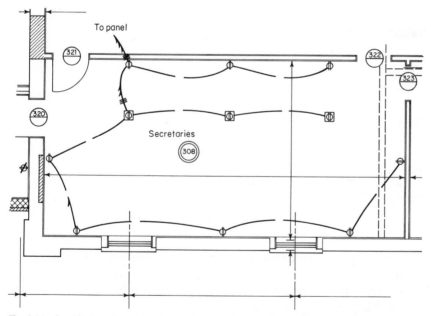

Fig. 1-21 Practical applications of receptacle symbols used on working drawings.

If other symbols are required to indicate various outlets on working drawings, they may be composed and added to the symbol list or legend with a description of their use. Examples are as follows:

 Duplex grounded receptacle with "Twist-Lock" connection

Whatever the need may be

Whatever the need may be

When outlets are located in areas requiring special boxes, covers, etc., they are usually indicated by abbreviations. Example:

W.P. Indicates weatherproof cover
E.P. Indicates explosionproof device, fittings, etc.
E or EMERG. Indicates outlet on emergency circuit

Service Equipment, Feeders, and Circuits

Panelboards, distribution centers, transformers, safety switches, and similar components of the electrical installation are indicated by electrical symbols on floor plans and by a combination of symbols and semipictorial drawings in riser diagrams. Some of these symbols are as follows:

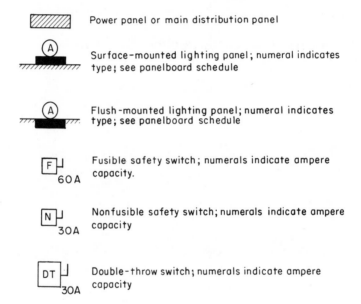

The description of panels and service equipment is usually covered in panelboard schedules such as those in Figs. 1-22 and 1-23; at other times a description of the panelboard is covered in the written specifications.

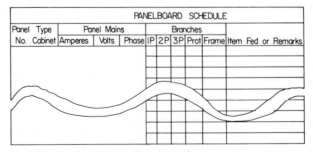

Fig. 1-22   Example of a panelboard schedule.

C.C.T No.	Volt - Amperes φ A	φ B	Description	Outlets L t g	R e c	O t l	T a	Cctbkr	Phase A B	Cctbkr T a	R e c	Outlets L t g	Description	Volt - Amperes φ A	φ B	C.C.T No.
1																2
3																4
5																6
7																8
9																10
11																12
13																14
15																16
17																18
33																34
35																36
37																38
39																40

Panel _____ _____ 1 & 3 Wire _____ Mounted _____ Ampere Main _____
Location _____ Ampere Bus _____

Subtotals

Total VA / φ
Total volt-amperes
Line amperes

Fig. 1-23 Another example of a panelboard schedule.

Figure 1-24 (*a* and *b*) shows how service equipment is shown on working drawings.

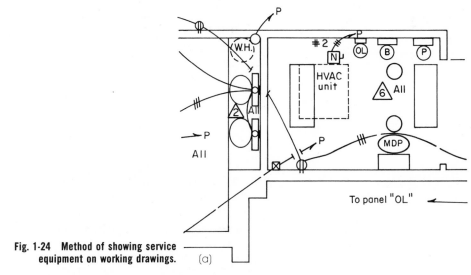

Fig. 1-24 Method of showing service equipment on working drawings. (a)

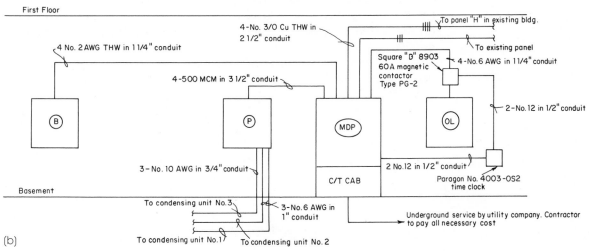

(b)

Circuit and feeder symbols have been nearly standardized in that most consulting firms use a solid line ——— to indicate circuits concealed in ceiling or wall and a broken line —— —— for conduit or raceways concealed in floor or ceiling below. Exposed conduit or raceway is also shown with a broken line, but the dashes are shorter than those used for concealed circuits: -----.

The variation between designers comes with the method of drawing these lines. Some designers prefer to draw all circuit lines with a straightedge, as shown in Fig. 1-25. Others prefer to use a French curve to draw the circuit line; this is illustrated in Fig. 1-26. Our firm prefers curved lines for all concealed circuits, for such lines are seldom, if ever, mistaken for building or equipment lines. However, on straight conduit wiring where conduit is run parallel with building lines, we do use straight lines to indicate such conduit, and the lines are drawn to show as closely as possible the actual route of installation.

We have seen some drawings using the symbol — X —— X —— X —— to indicate BX or flexible metallic armored cable in electrical systems where both rigid conduit and flexible cable were to be used.

The number of conductors in a conduit or raceway system may be indicated in the panelboard schedule under the appropriate column, but we prefer to show this information on the floor plans along with the circuits because workers have found this method to be the easiest to follow. For example, our symbol list contains the following symbol:

 # 10   Branch circuit concealed in ceiling or walls; slash marks indicate number of conductors in run—two conductors not noted; numerals indicate wire size—no. 12 AWG not noted

We are saying that a solid-line circuit with no slash marks or numerals indicates a circuit containing two No. 12 AWG conductors. Three slash marks with no numeral indicates three No. 12 AWG conductors, etc. If we want to indicate four No. 10 AWG conductors in a circuit, the symbol is ——// ——// ——.
10

Most electrical drawings use the symbol ——/// ➤➤ to indicate home runs to panelboard, with the number of arrowheads indicating the number of circuits in the run and the slash marks indicating the number of conductors in the run. However, since we use full arrowheads for call-outs as shown in Fig. 1-27, we use half-arrowheads to indicate branch-circuit home runs to panelboard: ———————/// ———————➤

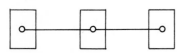

Fig. 1-25 Example of circuit lines drawn with a straightedge.

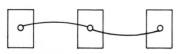

Fig. 1-26 Example showing circuit lines drawn with a French curve.

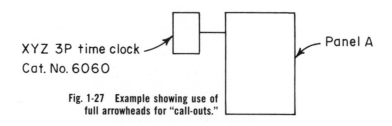

XYZ 3P time clock
Cat. No. 6060

Panel A

Fig. 1-27 Example showing use of full arrowheads for "call-outs."

Other circuits are indicated as follows:

——**T**——	Telephone conduit with pull wire
——**TV**——	Television-antenna conduit
——**E**——	Emergency circuit
—— - - ——	Low-voltage cable
——**EC**——	Empty conduit with pull wire

Communication and Alarm Symbols

Symbols for communication and signal systems, as well as symbols for light and power, should be drawn to an appropriate scale and accurately located with respect to the building in order to minimize the necessity to refer to architectural drawings. Where an extremely accurate final location of outlets and equipment is necessary, exact dimensions should be noted on the drawings.

If the communication and/or signal systems are highly complex, it is best to show these systems on drawings separate from the lighting and power systems. In any case, all circuits for communication, control, and signal systems should be laid out in complete detail on the electrical drawings, including identification of the number, size, and type of all conductors.

A complete communication and signal-system layout should include at least the following on one or more drawings:

1. Floor-plan layout, to scale, of all outlet and equipment locations, including routes of all circuits and wiring.

2. A complete schedule of all symbols used, with a clear description of each.

3. Where necessary for clarity, a single-line diagram showing the electrical relationship of the component items and sections of the wiring system.

4. Sections of the building or elevation of the structure may be required to clearly show outlet and equipment heights, relation to the established grade, general type of building construction, etc.

When the general wiring method and material requirements for the entire signal system are described in the specifications or by note on the drawings, no special notation need be made in relation to symbols on the drawing layout. For example, if an entire signal-system installation is required by the specifications and general references on the drawings to be "No. 14 AWG TW conductors; number as indicated in the wiring diagram; and installed in electrical metallic tubing (EMT) sized according to the National Electrical Code (NEC)," the outlet symbols do not need to have special identification. However, when certain different wiring methods or special materials will be required in different areas of the building or for certain outlets, such requirements should be clearly identified on the drawings by special identification or outlet symbols rather than only by reference in the specifications.

It has been recommended that each different basic category of signaling system be represented by a distinguishing basic symbol. Every item of equipment or outlet comprising that category of system is to be identified by that basic symbol. Then, different types of individual items of equipment or outlets indicated by a basic system symbol are further identified by a numeral or other identifying mark placed within the open-system basic symbol. All such individual symbols used on the drawings are included in the symbol list or legend.

The following (Fig. 1-28) is a list of signaling system outlets recommended by the American Institute of Electrical Engineers

Fig. 1-28 Signaling system outlets.

4.0 INSTITUTIONAL, COMMERCIAL, AND INDUSTRIAL OCCUPANCIES

Basic symbol	Examples of individual item identification (not a part of the Standard)	
4.1		I. Nurse-call system devices (any type)
	1	Nurses' annunciator (can add a number such as to indicate number of lamps)
	2	Call station, single-cord, pilot light
	3	Call station double-cord, microphonic speaker
	4	Corridor dome light. 1-lamp
	5	Transformer
		Any other item on same system—use numbers as required.
4.2		II Paging-system devices (any type)
	1	Keyboard
	2	Flush annunciator 2-Face annunciator
	3	Any other item on same system—use numbers as required. Flush annunciator
	4	2-Face annunciator
	5	Any other item on same system—use numbers as required.
4.3		III. Fire-alarm system devices (any type) including smoke- and sprinkler-alarm devices
	1	Control panel
	2	Station
	3	10" Gong
	4	Presignal chime
	5	Any other item on same system—use numbers as required.
4.4		IV. Staff-register system devices (any type)
	1	Phone-operators' register
	2	Entrance register—flush
	3	Staff-room register
	4	Transformer
	5	Any other item on same system—use numbers as required.
4.5		V. Electric-clock-system devices (any type)
	1	Master clock
	2	12" Secondary—flush
	3	12" Double dial—wall-mounted
	4	18" Skeleton dial
	5	Any other item on same system—use numbers as required.
4.6		VI. Public telephone system devices
	1	Switchboard
	2	Desk phone
	3	Any other item on same system—use numbers as required.

Basic symbol	Examples of individual item identification (not a part of the Standard)	
4.7		VII. Private telephone-system devices (any type)
	1	Switchboard
	2	Wall phone
		Any other item on same system—use numbers as required.
4.8		VIII. Watchman-system devices (any type)
	1	Central Station
	2	Key station
		Any other item on same system—use numbers as required.
4.9		IX. Sound system
	1	Amplifier
	2	Microphone
	3	Interior speaker
	4	Exterior speaker
	5	Any other item on same system—use numbers as required.
4.10		X. Other signal-system devices
	1	Buzzer
	2	Bell
	3	Push button
	4	Annunciator
	5	Any other item on same system—use numbers as required.

Signaling System Outlets

5.0 RESIDENTIAL OCCUPANCIES

Signaling system symbols for use in identifying standardized residential-type signal system items on residential drawings where a descriptive symbol list is not included on the drawing. When other signal system items are to be identified, use the above basic symbols for such items together with a descriptive symbol list.

5.1	•	Pushbutton
5.2		Buzzer
5.3		Bell
5.4		Combination Bell-buzzer
5.5	CH	Chime
5.6	◇	Annunciator
5.7	D	Electric Door Opener
5.8	M	Maid's signal plug
5.9	□	Interconnection box
5.10	BT	Bell-ringing transformer
5.11	▶	Outside telephone
5.12	▷	Interconnecting telephone
5.13	R	Radio outlet
5.14	TV	Television outlet

and published in 1962. While only basic signal-system outlet symbols are included in this list, the designer can readily improvise to fit any particular need which may arise.

Symbols used by our firm include the following with their respective descriptions.

Each year, electrical systems for new construction are augmented; new equipment and devices are introduced monthly which in turn require new electrical symbols for working drawings. Electric heat is being used more and more in new construction, and while forced-air systems are normally handled on the mechanical drawings, individual electric heating units are often specified to be supplied by the electrical contractor. We use the following symbols for electric heating equipment.

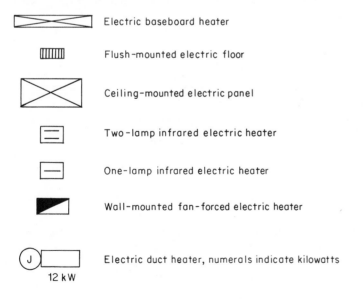

All the preceding symbols are drawn to approximate scale on the working drawings. An indicating mark identifies each unit, and a description of each appears in an electric heat schedule as shown in Fig. 1-29.

Htr. type	Manufacturer's description	Dimensions	Volts	Mounting	Remarks
①	Electromode CAT. No. 9955-H	104" in length	240	Baseboard	1,500 Watts
②	Electromode CAT. No. 9955-G	72" in length	240	Baseboard	1,000 Watts
③	Electromode CAT. No. 5990-W	18 1/4"h X 14 1/4"w	240	Recessed in wall	4,000 Watts
④	Electromode CAT. No. 8181-A	6" X 30" X 8"d	240	Recessed in floor	800 Watts
⑤	Electromode CAT. No. 9260/9261	22"l X 11"w X 3 1/2"d	240	Recessed in kickspace	2,000 Watts

ELECTRIC HEAT SCHEDULE

Fig. 1-29 Example of an electric heat schedule.

Ceiling or in-slab heat cables are indicated by the pictorial symbol:

All thermostats are shown by a circle with the letter T inside:

Electric heating contactors, when used, are indicated by a rectangle of the approximate scale size of the contactor cabinet; a descriptive note is then inserted to give an exact model number, etc., or it is indicated in the electric heat schedule and is given an indicating mark:

All circuits and feeders for electric heating equipment are shown in the conventional manner.

Surface metal raceway is also used a great deal on electrical drawings. The symbol for surface metal duct is:

W	W	W

If this metal duct contains outlets commonly known as Plug-mold, it is indicated by the symbol ▬■▬■▬■▬ with the shaded squares indicating the outlets. These outlets are also drawn to approximate scale to indicate the center-to-center distance.

Other such symbols include:

⊤T⊤ ⊤T⊤   Trolley duct

⊤B⊤ ⊤B⊤   Busway (service, feeder or plug-in)

⊤C⊤ ⊤C⊤   Cable trough

We use one common symbol for all motors regardless of their use. It consists of a circle with the motor horsepower inside:

⟨½⟩ . If the motor is enclosed inside a housing such as an air-handling unit, the approximate physical size and shape of the equipment are indicated by broken lines, and the motor is placed at the appropriate location inside the equipment, as shown in Fig. 1-30.

The Construction Specifications Institute, Inc., located at 1150 Seventeenth St., N.W., Washington, D.C. 20036, has just released Document no. 16015, "Electrical Reference Symbols." This document is a special publication of suggested abbreviations and symbols for use on engineering drawings and was prepared jointly with the Consulting Engineers Council/U.S. It is recommended that the reader obtain a copy of this document for reference.

Some electrical abbreviations include:

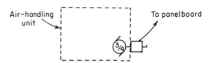

Fig. 1-30 Practical application showing motors on working drawings.

CSP	Central Switch Panel
DCP	Dimmer Control Panel
DT	Dusttight
ESP	Emergency Switch Panel
MT	Empty
EP	Explosionproof
G	Grounded
NL	Night Light
PC	Pull Chain
RT	Raintight
R	Recessed
XFER	Transfer
SFRMR	Transformer
VT	Vaportight
WT	Watertight
WP	Weatherproof

Chapter 2
Lighting with Incandescent Lamps

In the early days of electricity it became apparent that if a metal wire was heated by an electric current, it would glow and emit light. But such glowing metals—even platinum—rapidly deteriorated in air through oxidation. The problem was to find the proper combination of a filament and its environment.

As early as 1870, Thomas A. Edison used carbonized strips of paper in poorly evacuated glass envelopes to produce light with some degree of success. However, it was not until nine years later that he gave a practical demonstration of his lamps. Operating an electric generator in his laboratory, Edison provided light in the streets outside the laboratory as well as in a few of the adjacent houses.

Many great improvements in filament lamps were made in the years to follow, so that today such lamps have a much higher efficiency and longer life than anyone thought possible 50 years ago.

Incandescent lamps are now made in thousands of different types and colors from a fraction of a watt to over 10 kW each, and for practically any conceivable lighting application.

Regardless of the type or size, all incandescent filament lamps consist of a sealed glass envelope containing a filament. Light is produced when the filament is heated to incandescence (white light) by its resistance to a flow of electric current. Figure 2-1 illustrates the basic components of an incandescent lamp.

The quartz-iodine tungsten-filament lamp is basically an incandescent lamp, since light is produced from the incandescence of its coiled tungsten filament. However, the quartz lamp envelope is filled with an iodine vapor, which prevents the evaporation of the tungsten filament. This evaporation is what normally occurs in conventional incandescent lamps; then the bulb begins to blacken, light output deteriorates, and eventually the filament burns out. While the quartz-iodine lamp has approximately the same efficiency as an equivalent conventional incandescent lamp, it has the advantages of double the normal life, low lumen depreciation, and a smaller bulb for a given wattage. Figure 2-2 illustrates the basic components of a quartz-iodine lamp.

While fluorescent lighting has all but replaced incandescent lighting for general use in commercial buildings, incandescent fixtures are still the chief source of illumination for residences and for decorative lighting throughout the building construction industry. Incandescent fixtures are also very popular for localized or accent lighting in stores and other commercial interiors where special attention is demanded. The reason for their popularity is that incandescent lamps usually are more easily controlled than fluorescent or other electric-discharge lamps.

Projector and reflector incandescent lamps have proved useful for many special applications in the lighting of commercial buildings—especially for restaurants, motels, banks, various stores, and similar establishments. All such lamps, when carefully aimed to avoid creating annoying glare or reflections from mirrorlike surfaces, add greatly to the sales appeal, decor, and visual environment of the areas.

Colored reflector lamps are a convenient and effective source of colored light, especially when control of such light is needed. There are at least six colors available, all designed for maximum effectiveness when used either singly or in combination. For example, pink and blue-white lamps can be used to give an entire display a warm or cool tone with very little color distortion. Pink lamps compliment complexion tones; blue-white lamps are excellent for lighting displays of silverware, jewelry, or home appliances.

As mentioned previously, fluorescent lighting fixtures are the main source of light where a cool atmosphere or greatest economy is desired. A predominantly incandescent lighting system is used where a warmer atmosphere is desired or where color is of a critical nature, although deluxe warm-white fluorescent lamps give almost the same effect.

Most sales areas display high-profit or impulse items at the end of a counter or near the checkout area. These locations

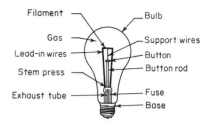

Fig. 2-1 Illustration showing the basic components of an incandescent lamp.

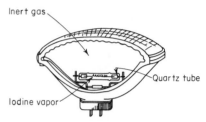

Fig. 2-2 Illustration showing the basic components of a quartz-iodine lamp.

should have from two to five times as much illumination as the rest of the counter, and incandescent spotlights with their accompanying highlights and shadows are very effective for this purpose.

Some of the advantages of incandescent lamps are as follows:

- Relatively low initial installation cost
- Not greatly affected by ambient temperatures
- Easily controlled as to direction
- Easily controlled as to brightness
- High color quality

Some of the disadvantages include:

- Less efficient than most electric-discharge lamps
- Higher operating cost per lumen
- More heat produced per lumen than electric-discharge lighting, causing the need for a larger air-conditioning system which in turn increases operating cost

RESIDENTIAL LIGHTING

With the exception of very large residences and tract development houses, the size of the average residential electrical system has not been large enough to allow for the expense of preparing complete electrical working drawings and specifications. Usually such electrical systems were laid out by either the architect—in the form of a sketchy outlet layout—or the electrician on the job, often only as the work progressed. However, many technical developments in residential electrical use—such as electric heat, greater use of electrical appliances, various electronic alarm systems, new lighting techniques, and electrical remote-control systems—have greatly expanded the demand and extended the complexity of today's residential electrical systems.

Architects are finding they need a greater engineering knowledge for the modern all-electric residence and now are seeking the services of consulting engineers. Since incandescent lamps are the chief sources of lighting in the home, residential lighting techniques will be discussed in this chapter. Other design techniques for residential electrical systems will be covered later.

The basic requirements for residential lighting are that it provide adequate light of the right quality and give proper attention to the appearance and artistic features of the lighting fixtures and the effect they produce.

Properly designed lighting is one of the greatest comforts and conveniences that any home owner can enjoy, and lighting should be considered as important as furniture placement, choice of draperies, color schemes, etc., because it is one of the most important features of decorating the home.

Residential lighting does not require elaborate calculations, but until designers have acquired the necessary experience, they should use some guide for the initial planning. The following method has been recommended by General Electric. In order to understand this procedure, the reader should be familiar with the following definitions:

Lumen: a unit of light quantity (luminous flux) produced by a light source.

Footcandle: the amount of direct illumination on a surface 1 ft from the flame of a standard candle.

One lumen per square foot equals one **footcandle.**

It is important to remember, when calculating the total lumen requirements by this method, that lighter colors reflect light while darker colors absorb it. Therefore, to achieve the same lighting level, a room with dark surfaces will require a greater number of lumens than a room with lighter surfaces.

The following table lists the required lumens per square foot for various areas in the home. These are recommended lumens when portable lamps, surface-mounted fixtures, and structural-lighting techniques are used. When a high percentage of the light for the area comes from recessed lighting fixtures, the figures in the table should be approximately doubled.

LUMENS REQUIRED PER SQUARE FOOT

Living room	80	Bedroom	70
Dining room	45	Hallway	45
Kitchen	80	Laundry	70
Bathroom	65	Work bench area	70

SAMPLE PROBLEM

Determine the total lumens required to achieve 80 lm per sq ft in a living room that is 15 ft wide and 20 ft long.

SOLUTION

a Multiply room width by room length to find square footage

15 ft × 20 ft = 300 sq ft

b Multiply square footage by desired lumens per square foot to get total lumens

300 sq ft × 80 lm per sq ft = 24,000 lm

c Select the proper bulbs and/or tubes from the Lamp Lumen Output Table on page 31 to total 24,000 lm and decide where and how they will be used.

A typical room lighting plan based on a 24,000-lm total is as follows:

1. Four 40-W WWX (warm-white deluxe) fluorescents, at 2,080 lm each, installed in a drapery cornice.

2. One 200-W inside-frosted bulb, at 3,940 lm, used in a study lamp.

3. Two 75-W recessed R-30 spotlights, at 860 lm each, over the piano.

4. Two three-way (100–200–300-W) soft-white bulbs, at 4,730 lm each, in two table lamps.

5. One (50–100–150-W) soft-white three-way bulb, at 2,190 lm, in a chairside table lamp.

The total of all lamps used yields 25,630 lm, or 85 lm per sq ft. While this is slightly over the "desired" illumination level, it should be pointed out that this lumen method should be used only as a guide and not as an absolute rule. A slight differential between "desired" and "actual" lumens is thus permitted.

This lumen method of calculating residential lighting can be an important aid in planning the lighting for a single room or for an entire home. With this guide, it is possible to determine the "total light output" of all light sources in a given area. This makes it possible for a lighting designer to determine quickly the number of lumens required in a specific lighting design to achieve the desired illumination level.

PRACTICAL APPLICATION

The floor plan of a small residence is illustrated in Fig. 2-3 and serves to demonstrate how the lighting system of a residence is laid out.

Vestibule

The vestibule or entrance hall is the area which gives visitors their first impression of the house. Here the owners greet their guests and help to remove their coats. Carefully planned

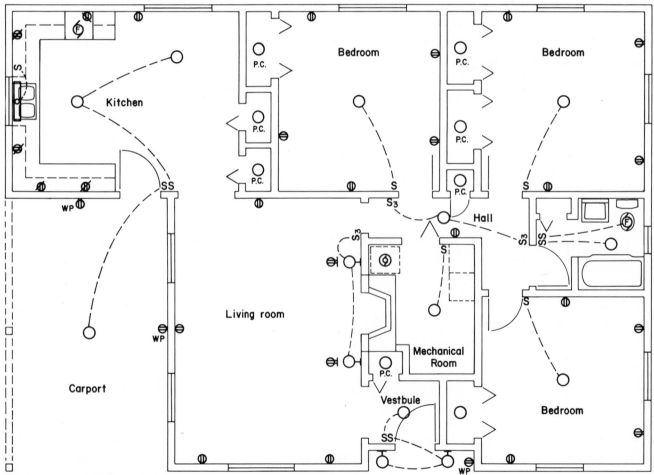

Fig. 2-3 Floor plan of a small residence.

LAMP LUMEN OUTPUT TABLE

Incandescent

Watts	Bulb	Designation and finish	Approx. initial lumens
25	A-19	Inside Frosted	232
		Soft-White	222
40	A-19	Inside Frosted	450
		Soft-White	435
		Dawn Pink	340
50	A-19	Inside Frosted	680
60	A-19	Inside Frosted	855
		Soft-White	840
		Dawn Pink	650
75	A-19	Inside Frosted	1170
		Soft-White	1140
		Dawn Pink	870
		Sky Blue	450
100	A-19	Inside Frosted	1750
		Soft-White	1710
100	A-21	Dawn Pink	1200
		Sky Blue	610
150	A-21	Inside Frosted	2830
		Soft-White	2710
150	R-40	Soft-White	2300
200	A-23	Inside Frosted	3940
		Soft-White	3840

Clear spots and floodlights

Watts	Finish and beam type	Bulb designation	Approx. initial lumens
30	Spot	R-20	200
50	Spot	R-20	430
75	Spot or Flood	PAR-38	745
75	Spot or Flood	R-30	860
150	Spot or Flood	PAR-38	1730
150	Spot or Flood	R-40	1950

Colored spots and floodlights

Percent initial lumen output—colored sources as relates to corresponding clear floodlights.

Watts and bulb designation	Amber	Blue	Blue white	Green	Pink	Red	Yellow
50wR20	—	—	55	—	74	—	—
75wR30/ 150wR40	35	10	30	15	60	15	95
100wPAR38	57	5	39	17	52	7	77
150wPAR38 DICHRO SP/FL	52	6	—	18	—	27	78

NOTE: Flair Chandelight as well as colored light sources are generally for decorative lighting and would not be included in a total lumen count.

3-Way bulbs

30/70/100	A-21	Soft-White	275/1010/1285
50/100/150	A-23	Soft-White	560/1630/2190
		Dawn Pink (med. base)	435/1253/1688
50/100/150	R-40	Soft-White Indirect	560/1630/2190
50/200/250	PS-25	Soft-White	550/3560/4110
100/200/300	PS-25	Soft-White	1290/3440/4730
		Dawn Pink (mogul base)	968/2580/3548

Fluorescent

Deluxe warm- or cool-white fluorescent			
Watts	Identification	Tube designation, thickness and length	Approx. initial lumens
14	WWX or CWX	T-12 (1½″ × 15″)	460
15	WWX or CWX	T-8 (1″ × 18″)	600
15	WWX or CWX	T-12 (1½″ × 18″)	505
20	WWX or CWX	T-12 (1½″ × 24)	820
30	WWX or CWX	T-8 (1″ × 36″)	1520
30	WWX or CWX	T-12 (1½″ × 36″)	1480
40	WWX or CWX	T-12 (1½″ × 48″)	2080

Deluxe warm- or cool-white circline fluorescent			
Watts	Identification	Tube designation, thickness and diameter	Approx. initial lumens
22	WWX or CWX	T-9 (1⅛″ × 8¼″)	745
32	WWX or CWX	T-10 (1¼″ × 12″)	1240
40	WWX or CWX	T-10 (1¼″ × 16″)	1760

NOTE: Deluxe white fluorescents render colors as they really are rather than distorting. Deluxe warm-white nearly duplicates the color of incandescent light while deluxe cool-white closely resembles daylight.

SOURCE: The General Electric Company.

lighting will show it to best advantage and will help the owners impart graciousness and hospitality.

By scaling the vestibule in Fig. 2-3, it is found that the dimen-

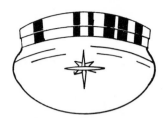

Fig. 2-4 Illustration of a close-to-ceiling lighting fixture.

sions are 4 ft 10 in. × 4 ft 4 in., or approximately 21 sq ft. According to the table on page 29, the recommendation for a hallway is 45 lm per sq ft. Thus, 21 sq ft × 45 lm per sq ft = 945 lm of lighting required in this area.

The table indicates that a 75-W inside-frosted A-19 lamp produces 1,170 lm (this is close enough for our purposes). The designer then selects a close-to-ceiling lighting fixture with diffusing glass, as shown in Fig. 2-4. This unit is lamped with the 75-W inside-frosted bulb.

Living Room

The living room is, of course, one of the most important rooms to have well lighted, since in the average home, which is the one under consideration, this room is the one in which members of the family spend much of their time, and also one that the owners wish to have most attractive when guests are present.

Referring again to the lumen method and to the drawing in Fig. 2-3, we find that the living room scales to 14 ft 0 in. × 16 ft 0 in., or 224 sq ft. The square footage is then multiplied by 80, the number of lumens per square foot recommended in the table.

224 sq ft × 80 lm per sq ft = 17,920 lm required

The owners expressed their desire to have mostly table lamps in the living areas which would provide local or functional lighting for visual tasks. Enough duplex receptables were placed around the perimeter to furnish electricity for these table lamps, which were to include: two (100–200–300-W) soft-white three-way bulbs, at 4,730 lm each, in two large table lamps at each end of the sofa; and one (50–100–150-W) soft-white three-way bulb, at 2,190 lm, in another table lamp placed next to a reclining chair. The three table lamps produced a total of 11,650 lm.

In order to achieve the desired 17,920 lm, the designer then chose two wall-mounted fixtures to be placed on each side of the fireplace. Each of these fixtures contained two 100-W A-19 inside-frosted lamps, at 1,750 lm each, or a total of 7,000 lm. Together with the table lamps, a total of 18,650 lm was available for lighting in the living area. A rheostat dimmer controlled the two wall fixtures for added versatility.

Bedroom Lighting

Bedrooms should be lighted with soft light that is not tiring to the eyes of a person lying in bed. Ceiling fixtures of the type shown in Fig. 2-5 are very good.

Fig. 2-5 One type of bedroom ceiling fixture.

It is also very important to have sufficient light at dressing tables and on mirrors; either wall-mounted lamps or vanity-table lamps should be provided on each side of the mirror.

Portable lamps on small tables by the beds or clamp lights to mount on the headboards are ideal for reading. A lamp at one side of the bed should direct its light so as to provide adequate reading light for the person on that side of the bed, yet not greatly disturb the person on the other side.

A switch controlling one of the lights in the room should be located near enough to a bed to be within easy reach of a person either in the bed or right at its edge. A table lamp with a built-in switch will, of course, accomplish this.

All the bedrooms in the floor plan in Fig. 2-3 are approximately the same size, that is, 158 sq ft. Referring to the chart, we find that bedrooms require 70 lm per sq ft. Therefore,

70 lm per sq ft × 158 sq ft = 11,060 lm required

The surface-mounted ceiling fixtures installed in the center of each room contain three 75-W inside-frosted lamps, at 1,170 lm each, for a total of 3,510 lm. The remaining lumens (7,550) desired were obtained by using table lamps at each side of the bed, each containing one (50–100–150-W) soft-white three-way bulb at 2,190 lm, and two more table lamps on the dresser with a 100-W inside-frosted lamp in each (1,750 lm apiece). This gave a total of 11,390 lm of illumination in the bedroom.

Hall

Since the hallway in the residence under consideration has a total of 36 sq ft, the total of lumens required is 36 × 45 = 1,620. A lighting fixture similar to the one in the vestibule is used, except that this time one 100-W A-19 inside-frosted lamp is used, and this produces 1,750 lm.

Mechanical Room

The mechanical room in this residence contains the furnace and water heater, also a washer and dryer. Therefore, this room is treated as a laundry area requiring 70 lm per sq ft.

60 sq ft × 70 lm per sq ft = 4,200 lm

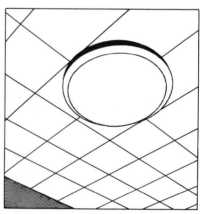

Fig. 2-6 One type of lighting fixture recommended for mechanical or utility rooms.

A lighting fixture was chosen that contained four 75-W inside-frosted lamps at 1,170 lm each. This fixture is shown in Fig. 2-6, and the diffusing shade gives a relatively soft and comfortable light for such a high intensity of illumination. The fixture produced 4,680 lm of light.

Kitchen

The kitchen should always receive careful attention, because it is an area where the homemaker spends a great deal of time.

Good kitchen lighting begins with a light source located in the center of the room and close to the ceiling. Since this is the fixture to furnish general illumination, it should be a glarefree source which will direct light to every corner in the kitchen. The homemaker will find work a lot easier if certain "task" areas are specially lighted — for example, counter lights beneath cabinets, or, as shown in our house plan, a recessed fixture over the sink to provide good supplementary lighting and eliminate shadows.

The kitchen-dinette shown in the floor plan has a pulldown fixture directly over the table for functional, yet decorative illumination. A combination hood/fan/light is also installed over the kitchen range, producing supplementary lighting for cooking. The two rooms are treated as one area needing 80 lm per sq ft.

80 lm per sq ft × 219 sq ft = 17,520 lm

Recessed fixture over sink	875 lm
Surface-mounted fixtures in center of kitchen area	4,680 lm
Range light	2,340 lm
Dinette light	4,730 lm
Total	12,625 lm

The owners did not want to add any additional lights at this point for reasons of economy. However, three 30-W under-cabinet lighting fixtures would have brought the total lumens to the desired 17,520.

Bathroom

Bathrooms should have either two wall-mounted fixtures, one on each side of the mirror, or one fixture mounted above the mirror for general light inside the room as well as for grooming. The light should be sufficient to illuminate one's face and the underside of the chin for shaving.

In the example of the house plan, one 20-W fluorescent fixture was used over the mirror to produce 820 lm. Since this was somewhat short of the 2,275-lm requirement determined previously, another lighting fixture was installed in the ceiling; this fixture contained one 100-W inside-frosted lamp at 1,750 lm. A similar fixture, containing one 60-W inside-frosted lamp, was placed in the area containing the linen closet.

OUTDOOR LIGHTING

Outdoor lighting is a partner in modern living. It welcomes guests and lights their paths, creating a hospitable look. As a safety factor, it protects the home from prowlers and reduces the incidence of outdoor accidents.

A pair of outdoor wall brackets, such as those shown at the front entrance on the floor plan in Fig. 2-3, is basic. If a long walk is necessary, a decorative yet practical post light should be included.

The carport is a separate lighting area, and, as shown in the floor plan, an all-purpose lighting fixture centered in the ceiling is adequate to light the owner's way to the back door.

RESIDENTIAL LIGHTING SUMMARY

The examples just given are basic for lighting a house. Obviously, no elaborate lighting effects were incorporated into the residence in Fig. 2-3 owing to the size and simplicity of the design. A larger or more elaborate house, such as that shown in Figs. 2-7 and 2-8, warrants added lighting effects, which may include:

1. Wall-wash fixtures except lights to dramatize fireplaces and bring out the texture of the bricks
2. Supplementary accent lighting over planters as well as for paintings or other pieces of art
3. Additional supplementary lighting around windows or in valances, or cove effects with hidden fluorescent lighting
4. Pendant-hung fixtures over bar areas
5. Luminous ceilings in kitchen and bathrooms
6. "Dressing-room" lighting around bath and vanity mirrors
7. Decorative yet functional lights in game and hobby areas
8. Swimming-pool lighting
9. Outdoor lighting which turns the patio and yard into an extra living or recreation room at night
10. For decorative and safety reasons, recessed or "bullet" lights installed in soffits at house corners

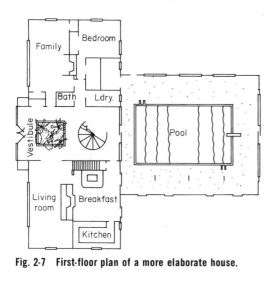

Fig. 2-7   First-floor plan of a more elaborate house.

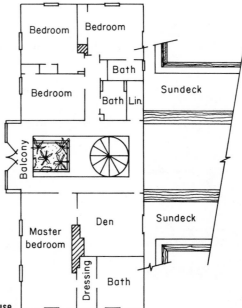

Fig. 2-8   Second-floor plan of a more elaborate house.

The varieties are endless, and lighting designers should use their experience and judgment to give the very best lighting system to fit the home owner's needs and budget.

COMMERCIAL LIGHTING

The floor plan of a branch bank in Fig. 2-9 illustrates a few applications of incandescent lighting fixtures in commercial buildings. While the general lighting is provided by recessed fluorescent fixtures, various effects could not have been accomplished without the use of incandescent lamps.

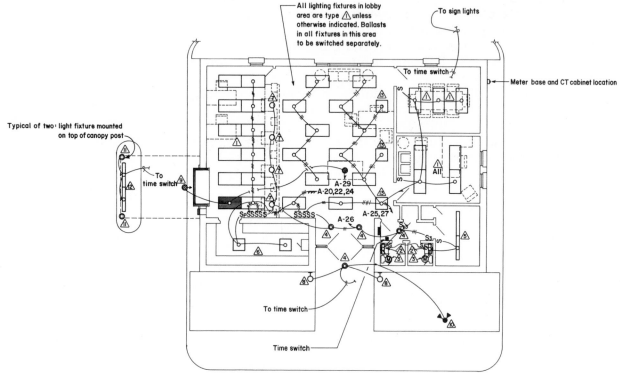

Fig. 2-9   Floor plan of a branch-bank building.

For example, the chandelier shown in Fig. 2-10 (without lamps) provides a decorative fixture during office hours and serves as a night light during closing periods.

Fig. 2-10   Chandelier used in lobby of branch bank.

Fig. 2-11   Pendant cylinders used over each teller area in the branch bank.

The pendant cylinders over each teller area are used as decorative fixtures as well as supplementary lighting for teller functions. See Fig. 2-11 at the left.

Recessed downlights are used in the vestibule to provide adequate lighting where the ambient temperature might be too low during cold weather to allow fluorescent fixtures to start properly. See Fig. 2-12.

The outside wall brackets shown in Fig. 2-13 contain 100-W incandescent lamps inside a frosted chimney to provide a Colonial feeling and blend with the Williamsburg style of the building.

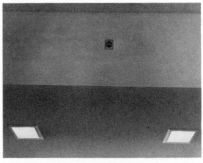

Fig. 2-12 Recessed downlights used in the vestibule of the branch bank.

Fig. 2-13 Wall-bracket fixtures used on the outside of the branch bank.

The inexpensive ground-mounted floor lights shown in Fig. 2-14 are used to illuminate the sign mounted on a vertical building wall.

Other examples of the use of commercial incandescent lamps are described in Chap. 5.

Fig. 2-14 An inexpensive ground-mounted floodlight used to illuminate a sign mounted on a vertical building wall.

Chapter 3
Lighting with Fluorescent Lamps

Since fluorescent lighting was introduced to the general public during the Chicago Centennial Exposition of 1933, it has become the major light source for the general interior lighting of commercial and institutional buildings, and during recent years, it has challenged other sources of light for use in residential, exterior, and other lighting applications.

One of the reasons for the popularity of fluorescent lighting is its high efficiency as compared to incandescent lamps. The average 40-W inside-frosted (IF) incandescent lamp delivers approximately 470 initial lm, while a cool-white fluorescent lamp of the same wattage delivers over 3,100 lm. This power efficiency not only saves on the cost of power consumed, but also lessens the heat produced by lamps which in turn reduces air-conditioning loads (another saving in power) and makes it more comfortable for those working under bright lights during warm weather.

Fluorescent lamps are available in straight, U-shaped, and circular configurations of various diameters. The insides of these glass-tube configurations are coated with a highly sensitive fluorescent powder (phosphor) which is activated by ultraviolet radiation and, in turn, converts the invisible energy to visible light. Each phosphor produces a characteristic visible light. Variation in the mixtures of various phosphors offers lamps in a wide range of visible-light colors. The most commonly used are as follows.

Cool White

This lamp is often selected for offices, factories, and commercial areas where a psychologically cool working atmosphere is desirable. This is the most popular of all fluorescent lamp colors since it gives a natural outdoor-lighting effect and is one of the most efficient fluorescent lamps manufactured today.

Deluxe Cool White

This lamp is used for the same general applications as the standard cool white, but contains more red, which emphasizes pink skin tones and is therefore more flattering to the appearance of people. Deluxe cool white also gives a good appearance to lean meat, keeps fats white, and emphasizes the fresh, crisp appearance of green vegetables. This type of lamp is generally chosen wherever very uniform color rendition is desired. However, it is less efficient than cool white.

Standard Warm White

Whenever a warm social atmosphere is desirable in areas that are not color-critical, this color is used. It approaches incandescent lamps in color and is appropriate whenever a mixture of fluorescent and incandescent lamps is used. While it gives an acceptable appearance to people, it has some tendency to emphasize sallowness. Yellow, orange, and tan interior finishes are emphasized by the standard warm-white lamp, and its beige tint gives a bright, warm appearance to reds, brings out the yellow in green, and adds a warm tone to blue. It imparts a yellowish white or yellowish gray appearance to neutral surfaces.

Deluxe Warm White

Deluxe warm-white lamps are more flattering to complexions than the standard and are very similar to incandescent lamps in that they impart a ruddy or tanned hue to the skin. They are generally recommended for applications in homes or social environments and for commercial use where flattering effects on people and merchandise are considered important. This type of lamp enhances the appearance of poultry, cheese, and baked goods. Deluxe warm-white lamps are approximately 25 percent less efficient than standard warm-white lamps.

White

White lamps are used for general lighting applications in offices, schools, stores, and homes where a cool, working atmosphere or warm, social atmosphere is not critical. They emphasize yellow, yellow-green, and orange interior finishes.

Daylight

In industry and work areas where the blue color associated with the "north light" of actual daylight is preferred, this lamp is often used. While it makes blues and greens bright and clear, it tends to tone down reds, oranges, and yellows.

In general, the designations "warm" and "cool" represent the differences between artificial light and natural daylight in the feeling they impart to an area. Their deluxe counterparts have a greater amount of red light, supplied by a second phosphor within the tube. The red light shows colors more naturally, but at a sacrifice of efficiency.

Other colors of fluorescent lamps are available in sizes that make them interchangeable with white lamps. These colored lamps are best used for flooding large areas with colored light; where a colored light of small area must be projected at a distant object, incandescent lamps using colored filters are best.

PRACTICAL APPLICATION: THE FIRST NATIONAL BANK, BRANCH BANK

From preliminary conferences with the architect and owners of the building the electrical designer has determined that the lighting system for this building should provide adequate lighting levels for the various areas, must complement the architectural design, and must fit within a predetermined budget. With these data in hand the electrical designer begins her or his design.

First, lighting calculations are begun by using the form shown in Fig. 3-1. The project identification is entered in the appropriate space at the top of the form; the designer's initials and the date are also entered—at the bottom of the form. Since the office (see Fig. 3-3) will be the first area under consideration, this area designation is entered in the same space as the project identification.

A check through the IES *Lighting Handbook*, under "Recommended Levels of Illumination," reveals that an office of this type requires between 100 and 150 fc. This fact is then entered in the appropriate space on the form.

The architectural drawings indicate an inverted T-bar ceiling system with 2- by 4-ft lay-in accoustical-tile panels. From this the designer decides upon a 2 by 4 ft recessed-troffer lighting fixture for the light source in this area. The catalog data are then entered in the correct space, again as shown in Fig. 3-1. Note that this fixture is designed to handle four 40-W fluorescent lamps, and the designer decides upon cool-white lamps for reasons described earlier in the chapter. A manufacturer's lamp catalog gives the initial lumens of this type of lamp to be 3,150. Since the fixture contains four such lamps, simple multiplication (4 × 3,150) gives the total lumens for each fixture as 12,600. All these data are again entered on the form.

The maintenance for this building is assumed to be good and the factor (MF) will therefore be 0.75, as given in the lamp manufacturer's data in Fig. 3-2 on page 42.

Selection of coefficient of utilization requires five basic steps (Fig. 3-1) and begins by filling in the sketch on the form.

ILLUMINATION CALCULATION SHEET

For Use with the IES Zonal Cavity Method

General Information

Project Identification: _First National Bank - Office_

Average Maintained Illumination for Design: _150_ footcandles.

Luminaire Data:

Manufacturer: _Bestlite Corp._ Catalog Number:: _54321-A_

Lamps (type & color): _40 WF (cool white)_ Number per Luminaire: _4_

Total Lumens per Luminaire: _12,600_ Maintenance Factor: _75_

Selection of Coefficient of Utilization

Step 1: Fill in sketch at right.

Step 2: Determine cavity ratios by the formula:

$$\frac{5 \times h \begin{bmatrix} CC \\ RC \\ FC \end{bmatrix} \times (L+W)}{L \times W}$$

Room cavity ratio, RCR = _4.7 or 5_

Ceiling cavity ratio, CCR = _0_

Floor cavity ratio, FCR = _2.13_

L = _12.5'_

$\rho = $ _80%_

$\rho = $ _%_

W = _11'_ $\rho = $ _50%_ WORK PLANE

$\rho = $ _%_ $\rho = $ _10%_

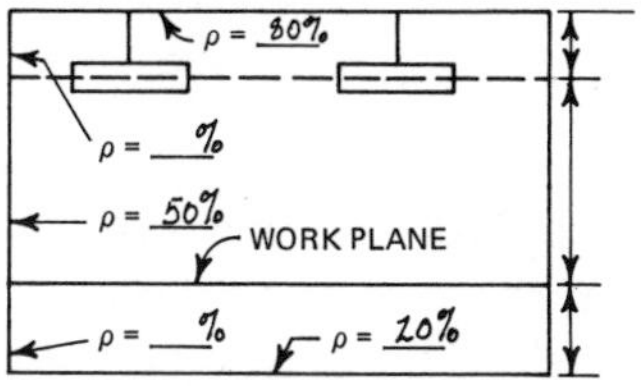

Step 3: Obtain effective ceiling cavity reflectance (ρCC). ρCC = _80%_

Step 4: Obtain effective floor cavity reflectance (ρFC). ρFC = _20%_

Step 5: Obtain coefficient of utilization (CU) from manufacturer's data. CU = _49%_

_4.42 or 4_

Calculations

Step 6:

Average Maintained Illumination Level

Footcandles = $\dfrac{\text{(Total lamp lumens per Luminaire)} \times \text{(CU)} \times \text{maintenance factor}}{\text{area per Luminaire}}$

= $\dfrac{12,600 \times .49 \times .75}{34.4}$

= _134.6_ footcandles on work area

Area per Luminaire: (This area divided by the Luminaire length gives the approximate spacing between continuous rows, or it may be divided into the total room area to determine the number of Luminaires required.)

Area per Luminaire = $\dfrac{\text{(total lamp lumens per Luminaire)} \times \text{(CU)} \times \text{maintenance factor}}{\text{footcandles}}$

= $\dfrac{12,600 \times .49 \times .75}{134.6}$

= _34.4_ square feet

CALCULATED BY: _JET_ DATE: _10/4/73_

Fig. 3-1 Lighting calculation form.

Step 1. (a) Room length (*L*) is found to be 12 ft 6 in. by scaling the architectural floor plan. Room width (*W*) is 11 ft 0 in. and is found by the same method.

(b) The ceiling, wall, and floor finishes are also found on the architectural drawings; then by checking the IES *Lighting Handbook*, the reflectances are as follows:

1. Ceiling, 80 percent
2. Walls, 50 percent
3. Floors 20 percent

Average luminance in footlamberts		
Angle from nadir	End-wise	Cross-wise
45	1286	1526
55	1011	1086
65	648	700
75	527	655
85	527	623

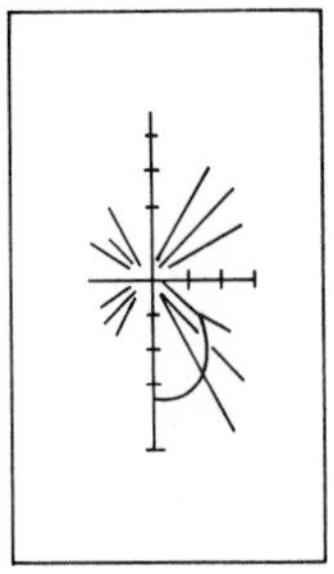

Distribution and efficiency
CIE % Luminaire
0
100

Spacing ratios and reflectance
Spacing not to exceed 1.4 x mounting height
Maintenance factors Good: .75 Med: .70 Poor: .65

	Effective ceiling cavity reflectance, R_{CC}								
	80%			50%			10%		
	% Wall reflectance, R_{CW}								
Room cavity ratio	50%	30%	10%	50%	30%	10%	50%	30%	10%
	Coefficients of utilization for 20% floor cavity reflectance, R_{FC}								
1	75	73	71	71	69	67	66	64	63
2	67	63	60	64	61	58	60	57	56
3	61	56	52	58	54	51	54	51	49
4	55	49	45	52	48	44	49	47	43
5	49	43	39	47	42	39	44	41	38
6	44	38	34	42	37	34	40	36	33
7	40	34	30	38	33	30	36	32	29
8	36	30	26	34	29	26	32	29	26
9	32	27	23	31	26	23	29	25	22
10	29	24	20	28	23	20	27	23	20

Fig. 3-2 Example of a lamp manufacturer's data sheet.

(c) The ceiling cavity ratio (CCR) is zero because the fixtures are recessed.

(d) The work plane is determined to be 30 in. (height of desk), or 2.5 ft above the floor.

(e) Since the total ceiling height has been determined from the architectural drawings to be 8 ft 0 in., the room cavity height is 5.5 ft (8 ft − 2.5 ft).

All these data are entered in the appropriate space (see Fig. 3-1).

Step 2. Determine Cavity Ratios (a) The room cavity ratio (RCR) is determined by substituting values in the formula given in Fig. 3-1. Thus,

$$\frac{5 \times 5.5 \, (12.5 + 11)}{12.5 \times 11} = 4.7$$

(b) The ceiling cavity is zero and requires no calculations.
(c) The floor cavity ratio is determined by substituting values in the formula used for (a). Thus,

$$\frac{5 \times 2.5 \, (12.5 + 11)}{12.5 \times 11} = 2.13$$

These values are also entered in the appropriate spaces on the form.

Step 3. Obtain Effective Ceiling Cavity Reflectance Since the CCR is zero, because of recessed fixtures the ceiling cavity reflectance is full value, or 80 percent in this case.

Step 4. Obtain Effective Floor Cavity Reflectance The floor cavity ratio has been calculated to be 2.13; therefore an adjustment may be necessary for the initial 20 percent floor reflectance. However, since there are many books available which explain the method of adjustment, we will not go into the method in detail at this point. Rather we will assume the floor reflectance to be the initial 20 percent.

Step 5.   Obtain Coefficient of Utilization   This is obtained from the manufacturer's data shown in Fig. 3-2. That is, in the column for 80 percent ceiling reflectance, 50 percent wall reflectance, 20 percent floor reflectance, and a RCR of 5 (4.7 rounded off), the coefficient of utilization is (CU) 49 percent.

Again all data are entered in the appropriate spaces on the form. With the data now complete, the number of lighting fixtures can be determined as well as the average illumination in footcandles.

Although the form in Fig. 3-1 does not have a space provided for calculating the number of fixtures required to produce the desired level of illumination, this will be the next step. The formula is as follows:

$$\text{No. of fixtures} = \frac{\text{floor area} \times \text{desired footcandles}}{\text{total lumens per fixture} \times \text{CU} \times \text{MF}}$$

Substituting known values in the formula, we have:

$$\text{No. of fixtures} = \frac{137.5 \times 150}{12,600 \times .49 \times .75} = 4.42$$

This figure (4.42), rounded off, indicates to the designer that four lighting fixtures should be used for this area. They are spaced as shown on the drawing in Fig. 3-3. It is now desirable to know the exact illumination of this area since the number of fixtures did not come out even. Proceed to step 6.

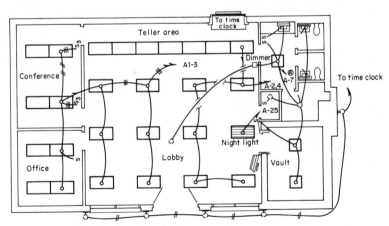

Fig. 3-3 Lighting diagram for bank office.

Step 6.   Substitute known values in the formula.

$$\text{FC} = \frac{12,600 \times .49 \times .75}{(137.5 \div 4) = 34.4} = 134.6 \text{ fc}$$

This satisfies the requirements of between 100 and 150 fc as recommended by the IES *Lighting Handbook.*

The remaining areas of this branch bank are designed in a similar manner in that the conference room was designed for approximately 125 fc, the lobby for 75 fc (using two-lamp fixtures), the teller area for 150+ fc, and the vault area about 50 fc.

Notice the fluorescent lighting fixture mounted on the outside of the building and over the drive-in window (Fig. 3-4). This fixture was provided to produce a level of illumination immediately outside the window that is approximately equal to the illumination level inside the bank. The reason is that during darkness, many drive-in-window tellers have found it difficult to see clearly out of the window because of interior reflections. This lighting fixture eliminates this problem.

After all the lighting fixtures have been laid out on the drawing, circuiting is the next step. In this building, all lighting branch circuits consist of No. 12 AWG TW conductors in ¾-in. conduit, protected with one-pole, 20-A circuit breakers installed in panel A. Therefore, the designer connects a lighting load of between 1.2 and 1.6 kW to each circuit. The electrical symbols given in Chap. 1 will clarify the meaning of each circuit.

The uses of the incandescent lighting fixtures were described in Chap. 2, and the power wiring, alarm system, etc., will be described later in this book.

The photographs (Figs. 3-4 to 3-7) are illustrations of the installed electrical system of the branch bank just described.

Fig. 3-4 Lighting fixture at a drive-in bank window.

Fig. 3-5 Office lighting fixtures installed in lay-in ceiling.

Fig. 3-6 Lighting in bank lobby and teller areas.

Fig. 3-7 Surface-mounted 2- by 2-ft lighting fixture in vault area.

On actual working drawings, a lighting fixture schedule and other details are usually provided on the drawing as shown in Fig. 3-3. However, due to limited space on the pages of this book, such details are not shown. A lighting fixture schedule such as that shown in Fig. 3-8 should be provided on all working drawings. This schedule should give all the information necessary to tell the contractors or their electricians exactly what fixture is to be installed, the type of mounting, number of lamps, etc.

LIGHTING FIXTURE SCHEDULE						
FIXT. Type	Manufacturer's Description	LAMPS No.	Type	Volts	Mounting	Remarks

Fig. 3-8 Lighting fixture schedule used for bank project.

PRACTICAL APPLICATION: LAUNDRY BUILDING

Figure 3-9 is the floor plan of a laundry building constructed to handle the in-house laundry for a district home for the aged. Since the building is of a utility class and a limited budget was provided for its construction, a very simple, low-cost lighting layout was designed.

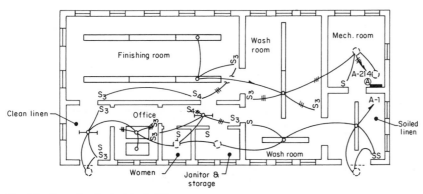

Fig. 3-9 Floor plan for laundry building.

Bare-lamp, 4-ft, surface-mounted lighting fixtures are used in the clean linen area and in the corridor. Each fixture contains two 40-W cool-white fluorescent lamps, and approximately 25 fc is provided in each area.

The office has two 18-in. by 4-ft fluorescent fixtures with acrylic prismatic diffusers which give the area over 150 fc of illumination.

Due to the fact that moisture is ever present in the remaining areas of the laundry, enclosed and gasketed 8-ft fluorescent fixtures are used. Each of these fixtures contains one 8-ft, 150-W fluorescent lamp and an 800-mA ballast. The acrylic clear prismatic lens on these fixtures provides for excellent light control; the prisms assure good brightness and hide lamp image. The continuous one-piece vinyl gasketing has no seams and is therefore dirt-, dust-, bug-, and watertight.

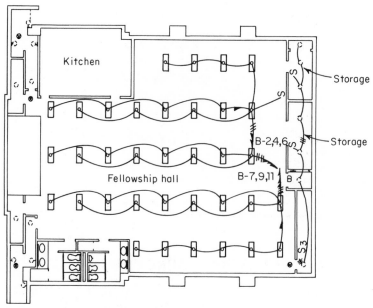

Fig. 3-10   Floor plan of existing lighting layout in church fellowship hall.

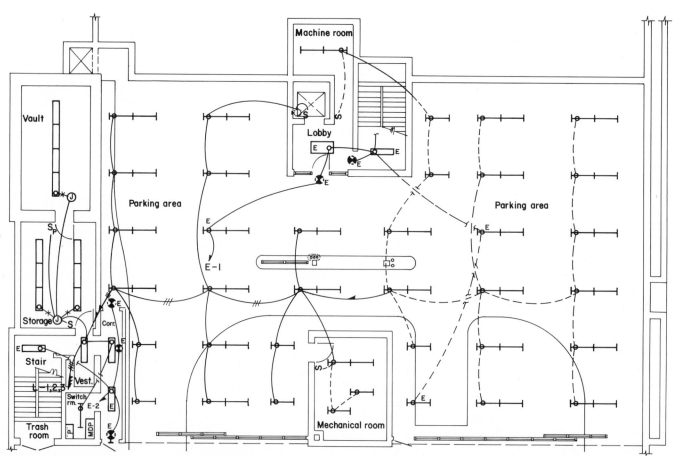

Fig. 3-11   Floor plan of lower floor of new bank building.

PRACTICAL APPLICATION: CHURCH BUILDING

Incandescent lighting fixtures were the principal light source for lighting the main worship area of the church building described in Chap. 1. The fellowship hall (located in the basement) of this same church is lighted with fluorescent fixtures as shown in Fig. 3-10. The general lighting source for this area consists of thirty-five 18- by 48-in. lighting fixtures containing four rapid-start 40-W fluorescent lamps. Acrylic prismatic diffusers are used throughout.

Notice that the fixtures are circuited so that several combinations of illumination are possible. For example, with all lamps energized, the illumination level is approximately 70 fc. Half of the fixtures may be lighted with the remaining half "out"; this gives approximately 40 fc. Although this is not shown on the floor plan in Fig. 3-10, the lamps in each fixture are controlled in pairs, giving further flexibility of lighting control.

The toilets are lighted with surface-mounted fluorescent strips with diffusers—two single-tube rapid-start 40-W fluorescent lamps in each area on the ceiling, and wall-mounted fluorescent fixtures over each mirror at the sinks.

Other sources of light for this area as well as the power wiring are described later in this book.

PRACTICAL APPLICATION:
THE FIRST NATIONAL BANK, MAIN OFFICE

Figure 3-11 shows the floor plan of the lower floor area of a new bank building. It is obvious that this area is mainly used for parking and a route to the vacuum-tube drive-in units. Other areas include storage room, vault, stairwells, and the mechanical and machine rooms.

The entire parking area is illuminated with surface-mounted bare-tube fluorescent strips—each 4 ft in length and containing one 40-W rapid-start fluorescent lamp. This lighting arrangement gives a minimum of 5 fc of illumination. Each fixture is also provided with wire lamp guards to protect them from automobile radio antennas.

Most of the other areas are lighted with surface-mounted fluorescent lighting fixtures containing two 40-W rapid-start lamps each. Diffusers are also provided for better appearance of the fixture itself and for greater visual comfort.

Most of the parking area lighting is controlled by circuit breakers within the lighting panel. However, five of these fixtures as well as two other fixtures are connected to the emergency panel for 24-hr lighting.

Figure 3-12 shows the plan of the main level of this same bank. The third level of this bank building (office level) is shown on the floor plan in Fig. 3-13. The book vault, women's lounge, and the toilets are lighted with surface-mounted, two-lamp, fluorescent fixtures with diffusing lenses. The corridor and stairwells are illuminated with single-tube fluorescent fixtures with diffused lenses. The remaining area on this floor is illuminated with 2- by 4-ft recessed troffers, each containing four 40-W fluorescent lamps.

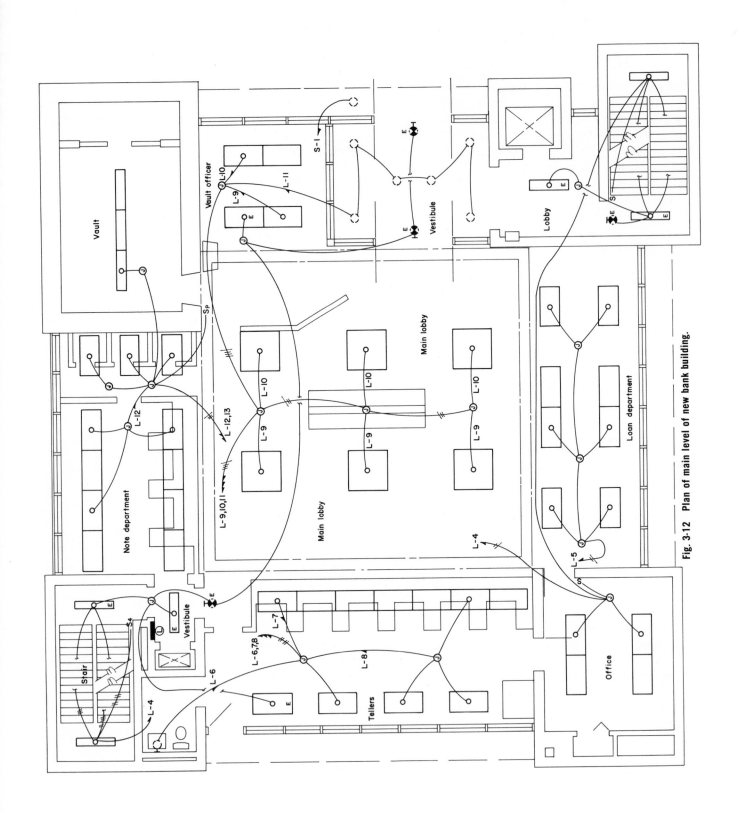

Fig. 3-12 Plan of main level of new bank building.

48

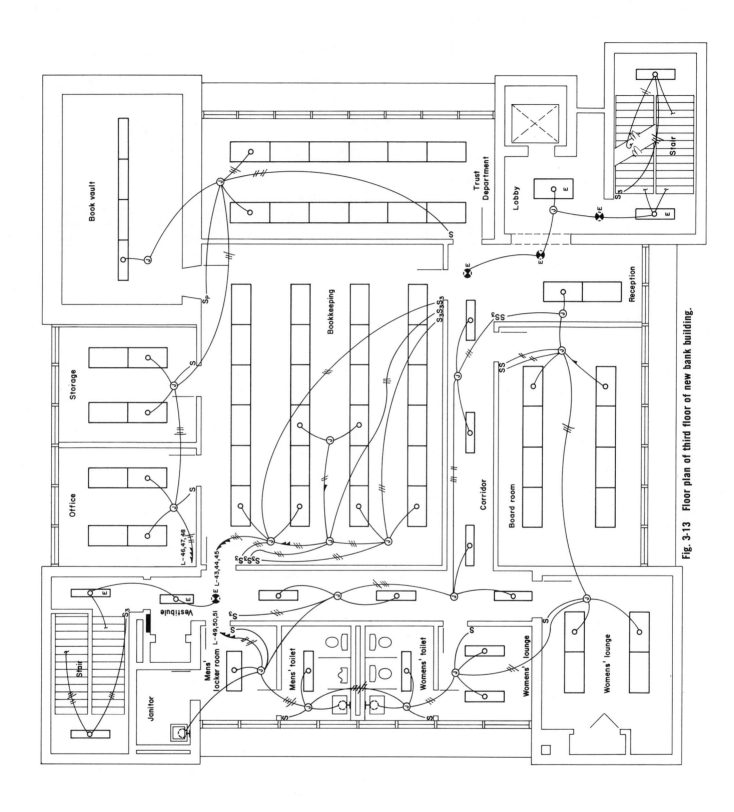

Fig. 3-13 Floor plan of third floor of new bank building.

49

Fig. 3-14 Four-foot-square fluorescent fixtures in bank lobby.

Fig. 3-15 Two-foot by four-foot recessed lighting fixtures in teller area.

The lobby (Fig. 3-14) is illuminated with six 4- by 4-ft surface-mounted fluorescent fixtures, each containing eight 40-W fluorescent lamps. This arrangement provides between 50 and 70 fc of illumination—the recommendation of the Illuminating Engineering Society (IES). The vault and corridors have surface-mounting fluorescent fixtures with diffusers, while the remaining lighting fixtures are four-tube, 2- by 4-ft recessed troffers with dual-prism lenses (Fig. 3-15).

Incandescent lamps used for highlighting the building are explained in Chap. 2. The power wiring, alarm systems, service, and feeders for this building are described in a later chapter. The explanation of electrical symbols was given in Chap. 1. The circuiting of these lighting circuits should be self-explanatory.

Other uses of fluorescent fixtures will be given in the chapter on outside lighting and also in the chapter on using combinations of incandescent and fluorescent lamps for varied effects (Chap. 5).

Chapter 4

Lighting with High-Intensity Discharge Lamps

ADVANTAGES AND DISADVANTAGES

The basic mercury vapor (MV) lamp produces light with a predominance of yellow and green rays combined with a small percentage of violet and blue. In light of this color small objects, such as screws, pins, nuts, bolts, etc., stand out very sharply; this effect can result in increased industrial production speed and fewer errors—with less eyestrain for employees. For the reasons just mentioned, the mercury lamp became highly popular for certain industrial applications—especially in automobile manufacturing plants and machine shops.

While a modern mercury lamp, when lighted, appears to emit a white light (red, blue, and green), red is absent; therefore red

51

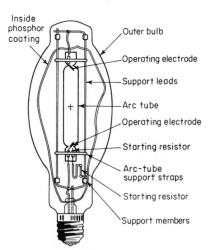

Inside phosphor coating

Outer bulb

Operating electrode

Support leads

Arc tube

Operating electrode

Starting resistor

Arc-tube support straps

Starting resistor

Support members

Fig. 4-1 Components of a typical mercury-vapor lamp.

objects appear black or dark brown under mercury vapor lamps. This color distortion has, in the past, prevented its use in many applications; however, it has now been overcome to a certain extent by the use of red light–generating chemicals within the bulb. Consequently, this lamp is now finding its way indoors for more and more commercial lighting applications.

Figure 4-1 shows typical mercury lamp components which include the following: an arc tube made of quartz to withstand the high temperatures resulting when the lamp builds up to normal wattage; two main operating electrodes located at opposite ends of the tube; a starting electrode connected in series with a starting resistor which is connected to the lead wire of the operating electrode; tube leads and supports; and an outer phosphor-coated bulb that helps to stabilize lamp operation and prevent oxidation of metal parts.

Mercury lamps for general-lighting purposes are available in wattages from 40 to 2000 W, in clear and various types of color-improved lamps. All types of mercury lamps require their own specially designed transformers, or ballasts, for proper starting and operating performance. While these ballasts are usually located outside the lamps, self-ballasted lamps with built-in filament-type ballasts are available.

One disadvantage of mercury lamps is that they are extinguished in the event of current interruption or excessively low voltages. They will not restart until they have cooled down and the internal vapor pressure has been reduced to the point of restarting the arc with the available voltage. The restarting-time interval is between four and seven minutes.

Besides being used to produce visible light, mercury lamps have been used to produce ultraviolet radiation as a source of "black light." This type of light requires that all visible light be screened out by filters while the ultraviolet energy passes through. Black-light lamps are used for decorative and theatrical effects, industrial inspection, medical and chemical analysis, criminal investigations, mineralogy, and varied military applications.

Metal halide lamps closely resemble regular clear mercury lamps, except that the inner arc tube in a metal halide lamp contains additional halide compounds to increase the light output and improve the lamp color. This improved color has made the metal halide lamp suitable for many indoor applications, including food displays in supermarkets. It has found more use, however, in outdoor floodlighting, sports lighting, and certain general street-lighting applications.

The high-pressure sodium lamp is another type of high-intensity discharge (HID) lamp in that it utilizes an arc tube to enclose gases through which an electric current passes. However, its unique light-transmitting ceramic tube enables sodium to be used at higher temperatures and pressures than were previously attainable. The result is a warm yellow light at nearly maximum theoretical efficiency, that is, 100 to 115 lm per watt.

Since some amount of every color is present in the high-pressure sodium lamp, it is used for practically all general lighting under most conditions.

Advantages of HID (Mercury) Lamps
- Long life.
- High lumen output.
- Compactness of incandescent lamps.
- Not affected by ambient temperatures as are regular fluorescent lamps.
- Better degree of light control than with fluorescent.

Disadvantages of HID (Mercury) Lamps
- Color acceptability low with clear mercury lamps.
- Sensitive to voltage variation.
- Long restarting time required.
- Light is extinguished if momentary current interruption occurs.
- Delay of between four and seven minutes from starting time to full brightness.

Obviously, the lighting demands of one area will be different than those of another, and since the available light sources have different performance characteristics, the demands of the application will dictate selection of the proper light source.

The following table illustrates the performance characteristics of the prevalent HID light sources.

Light source	Predominant color-spectrum properties	Normal life rating, hr	Initial luminous flux, lm/W	Beam control
MV (clear)	Blue	24,000	54	Good
MV (color improved)	Blue-green	24,000	54	Fair
MV (deluxe white)	Green-yellow	24,000	54	Fair
Metal halide additive	Green-yellow-red	7,500	80	Excellent
High-pressure sodium	Yellow	10,000	115	Excellent

DESIGN PROCEDURE

When lighting outdoor horizontal surfaces such as parking lots, storage lots, etc., the location, mounting height, and spacing of lighting fixtures are extremely important factors in attaining a uniform illumination on the surface. One rule of thumb is never to have a distance between pole locations of more than four times the mounting height of the fixtures, or not more than twice the mounting height from a central pole to the nearest perimeter.

This rule provides for the widest spacing permissible to obtain up to 10 fc of illumination on the surface. In areas which must receive higher levels of illumination, closer spacing between pole locations is necessary to accommodate the greater number of fixtures required. However, the reverse is not necessarily true. That is, if a lower level of illumination is desired, it would be better to use lamps of lower intensity than to increase the spacing of the lighting fixtures.

SAMPLE PROBLEM

A large parking lot (150 × 400 ft) is to be lighted to an average illumination level of 1 fc.

SOLUTION

Refer to the chart in Fig. 4-2 at 1.0 fc. Read horizontally to the right until this line intersects diagonal line *A*. Now read down vertically and notice that this line intersects at approximately 3.3—the initial lumens per square foot of area required to attain an average of 1 fc using a lighting fixture with a white translucent globe.

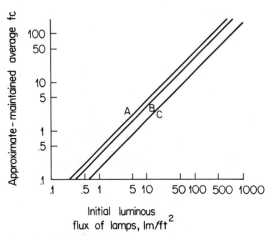

Fig. 4-2 Lighting calculation chart: *A:* clear globular lens; *B:* translucent globe; *C:* white ball globe.

Therefore

3.3 lm per sq ft × area = total lm required
3.3 lm per sq ft × 150 ft × 400 ft = 198,000 lm

A lamp catalog states that a 400-W mercury lamp provides 20,500 lm.

Therefore

$$\frac{198,000 \text{ lm required}}{20,500 \text{ lm per lamp}} = 9.65 \text{ or } 10 \text{ lamps}$$

Ten lighting fixtures and standards (poles) would be required to light this area, and they may be laid out as illustrated in Fig. 4-3.

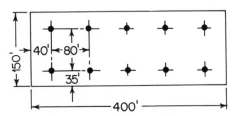

Fig. 4-3 Layout of parking-lot lighting fixtures.

Illumination uniformity is even more important when one is lighting such vertical surfaces as signs and billboards and in architectural lighting for vertical surfaces of buildings. After establishing footcandle levels required from the IES *Lighting Handbook*, the type and wattage of lamp necessary to produce these levels should be determined; the number of fixtures can then be calculated. Selection of lighting-fixture beam-spread and mounting distance to best achieve uniformity would be the next step.

The example in Fig. 4-4 demonstrates how medium-beam lighting fixtures, narrow-beam fixtures, and a combination of both types can be arranged to produce the most uniform beam distribution.

Lighting fixtures with various beam-spreads may be selected

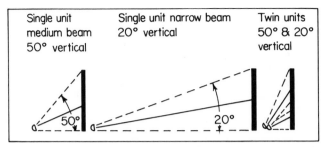

Fig. 4-4 Illustration of various types of lighting fixtures arranged to produce the most uniform beam distribution.

from manufacturers' catalogs. The best practice in selecting from available beam-spreads is to choose a lighting fixture which projects most of its effective beam-spread within the area to be illuminated with a minimum of "wasted" light (Fig. 4-5).

As the lamp lumens are distributed in a wider beam-spread, the candlepower decreases proportionately. Thus, the farther the light must be projected, the narrower the beam-spread should be. Narrow beam-spreads can also be used to advantage at closer ranges to provide concentrated high-intensity illumination levels.

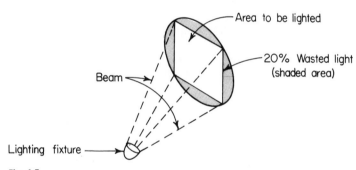

Fig. 4-5

Another point to remember is that projection-type lighting fixtures should never be mounted at an angle less than 30 degrees below horizontal. If this standard is maintained, the resultant glare will be minimized. In actual practice, if recommendations for pole spacing and lighting-fixture mounting height are followed, the aiming angle will automatically be greater than the minimum 30 degrees below horizontal. See Fig. 4-6.

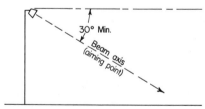

Fig. 4-6 Illustration showing aiming angle of projection-type lighting fixture.

PRACTICAL APPLICATION

The "watts per square foot" method is the easiest and most rapid procedure for calculating the number of lighting fixtures needed in most applications requiring general illumination levels of 10 fc or less. The formula is

$$\frac{\text{Area (sq ft)} \times \text{desired footcandles} \times \text{WSF}}{\text{Luminaire lamp watts}}$$

$$= \text{number of lighting fixtures}$$

where Area = square feet of surface to be illuminated

Footcandles = recommended illumination level for type of area to be illuminated

Luminaire lamp watts = total watts of lamp selected for given application

WSF = a utilization factor which combines lamp lumens, beam efficiency, and maintenance factor

It was determined that a single WSF factor would not give accurate results for areas of all sizes; however, the four factors in the following table will suffice for most areas as given.

WSF FACTOR	
Small area (1,000 to 3,000 sq ft)	0.16
Medium area (3,000 to 20,000 sq ft)	0.11
Large area (20,000 to 80,000 sq ft)	0.08
Extra-large area (over 80,000 sq ft)	0.06

For example, in the table above, the WSF factor for an extra-large area lighted with mercury vapor lamps is 0.06. This indicates that such an area requires 0.06 W of "lamp power" to place 1 fc of illumination within 1 sq ft of the area to be illuminated.

EXAMPLE: SHOPPING CENTER PARKING LOT

The site plan of a relatively small shopping center is shown in Fig. 4-7. The lighting designer was required to design a lighting

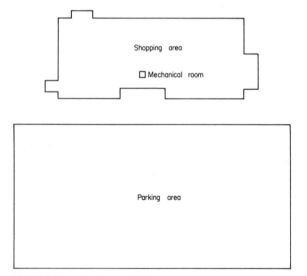

Fig. 4-7   Site plan of a small shopping center.

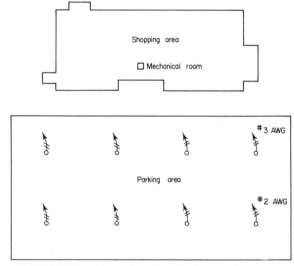

Fig. 4-8   Site plan, as shown in Fig. 4-7, with electrical components and circuits completed.

layout for the parking lot which had good appearance by day, gave efficient performance at night, and blended tastefully with the architectural design of the shopping center buildings. Illumination levels were to meet IES criteria.

SOLUTION

Step 1. Using the watts per square foot method as a basis, the designer first found the area of the parking lot by scaling the site plan drawings: $240 \times 480 = 115,200$ sq ft. Since the total area is over 80,000 sq ft, 0.06, the WSF factor for an extra-large area, must be used.

Step 2. The designer than referred to the table of recommended illumination levels in the IES *Lighting Handbook* and found the recommended level to be 2 to 5 fc for automobile parking areas in shopping centers. The designer chose 3 fc as a basis for design.

Step 3. Then 400-W lamps were chosen as the first trial lamps for use in the design. A large lamp catalog indicated that this type of lamp provided 20,500 initial lumens.

Step 4. Substituting known values in the WSF formula, we have

$$\frac{115,200 \times 3 \times 0.06}{400} = 51.84 \text{ or } 52 \text{ lamps}$$

Step 5. A judgment is now required of the designer. Fifty-two 400-W lamps would require 13 four-lamp fixtures and posts, a design found to be economically unwise. The designer then selects a lighting fixture which contains four lamps—two 400-W mercury lamps and two 1-kW mercury lamps. This of course requires two different sizes of ballast for each fixture.

Step 6. The new lamp wattages are entered in the WSF formula and recalculated.

$$\frac{115,200 \times 3 \times 0.06}{2(400 + 1,000)} = 7.41 \text{ or } 8$$

This makes an even number of fixtures.

Step 7. The designer then uses 8 four-lamp fixtures (2,800 W total per fixture) mounted on 30-ft aluminum standards and spaced 120 ft on centers. The location of the fixtures, as well as electrical circuits for each, is sketched and then given to an electrical draftsman for completion. The completed site plan drawing appears in Fig. 4-8. Meanwhile, the voltage drop must be considered before circuits can be sized. Control of these lighting fixtures must also be determined before the drawing will be complete.

Step 8. The owners' service-equipment room is located in one of the shopping-center buildings. The largest lighting-fixture circuit is approximately 370 ft distant from this room. Therefore, 370 ft will be used to calculate the voltage drop and, in turn, the required conductor size. It is desired to keep the voltage drop below 2 percent, i.e., 0.02×120 or 2.4 V.

Step 9. The total amperes per fixture is 2,800W/120 V = 23.3 A; the total length is 370 ft; the maximum voltage drop allowed is 2.4 V. The formula for determining wire size is, then,

$$\text{Circular mils} = \frac{370 \times 23.3 \times 19}{2.3} = 71,216$$

Reference to a circular mil wire-size chart shows that the closest wire size available to the 71,216 circular mils is No. 1 AWG copper. However, experience tells us that a No. 2 AWG conductor with an area of 66,370 circular mils will be sufficient, since the lighting fixtures will only operate an average of one or two hours per day.

The next fixture is 250 ft from the service equipment. Therefore, the conductor size for this fixture will be

$$\frac{250 \times 23.3 \times 19}{2.3} = 48,120 \text{ circular mils}$$

or No. 3 AWG conductor.

Each circuit is calculated in the same manner, and the appropriate wire size is indicated on the drawings. Since the total calculated load for each fixture is 23.3 A, all circuits for these parking-lot lighting fixtures are protected with a 30-A overcurrent device.

Step 10. Each lighting fixture is controlled by its own built-in photoelectric cell.

SUPERMARKET LIGHTING

It was mentioned earlier in this chapter that new HID lamp designs show a remarkable color improvement over earlier mercury lamps, and that they are now suitable for many indoor applications. One such application is the lighting system for the supermarket shown in Fig. 4-9.

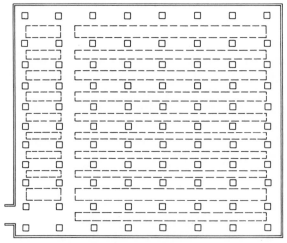

Fig. 4-9 Lighting system for a supermarket.

The required number of fixtures to produce the desired illumination level was calculated by using the zonal cavity method. Recessed HID lighting fixtures were used and were spaced 6 ft on center. The result of this design gave the ceiling a more spacious, clean, and nondistracting look. Hot spots and glare were kept to a minimum, while the acrylic refractors provided even illumination throughout the store area. Since the HID lamps provided more lumens per watt than incandescent or fluorescent fixtures, much energy was saved in both illumination and air-conditioning use.

INDUSTRIAL APPLICATIONS

Most industrial lighting applications are not color-critical, and all the HID light sources are usually satisfactory. The following suggestions for choosing the proper industrial light sources were furnished by General Electric.

1. Use Lucalox (high-pressure sodium) lamps where minimum cost of light is desired.

2. Use Multi-Vapor (tandem fixtures) where low cost of light and a cooler-appearing environment are desired.

3. Use semireflector deluxe white lamps in practically all mercury lamp applications. Some incandescent or fluorescent lighting fixtures should be added for standby lighting in case of a power interruption. Some HID lighting fixtures can be purchased with optional built-in quartz lamps connected to the same circuit which can be used for standby lighting.

4. Use silver-reflector mercury lamps in combination with silver-reflector filament lamps in areas with severe dirt conditions.

The lighting system illustrated in Fig. 4-10 consists of 400-W mercury fixtures mounted 10 × 10 ft on centers and produces approximately 100 fc of illumination.

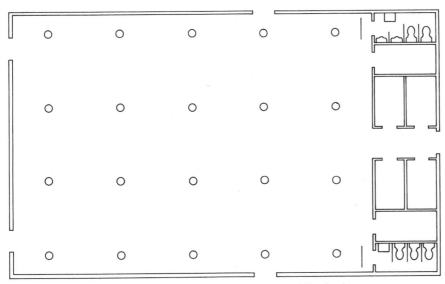

Fig. 4-10 Industrial lighting system designed to produce 100 fc of illumination.

Chapter 5

Lighting with Combinations

While you were reading Chaps. 1 through 4, it probably became apparent to you that modern lighting techniques, combined with new ideas of interior architectural design, have greatly influenced the appearance of interior environments.

When we consider interior lighting practice in the various fields of application, we are inclined to think, first of all, of the general overall lighting of a room and of the various types of lamps and lighting fixtures employed. In many interior lighting applications, however, other specialized lighting facilities may be just as important, if not more so. For such facilities, the use of a combination of different types of lamps with their related lighting fixtures offers an endless number of moods and effects.

For critical color matching, no one light source is completely satisfactory, since colors which match under one light source may not match under another. Consequently, for the most accurate work, two sources having different light-color characteristics are used. For example, a fluorescent lighting fixture using "daylight-color" lamps in combination with an incandescent lamp and fixture offers widely different colors and should be satisfactory for critical color matching for most items. However, both light sources should be well diffused.

Light-source combinations, such as a combination of fluorescent and incandescent lamps, are also widely used in most stores, to achieve a particular objective. One example would be the use of incandescent downlights alternated between sections of fluorescent fixtures, or an arrangement of eyeball spotlights at each end of an 8-ft fluorescent fixture.

A combination of different light sources adds greatly to the appearance of merchandise. A deep-pile rug, for example, has a textured finish. Evenly diffused fluorescent light may illuminate each fiber of such a rug so uniformly that it would be difficult to appraise the pattern and depth of the pile. However, if directional incandescent light is used in combination with diffused fluorescent light, the soft texture and deep nap of the rug will be greatly emphasized.

Shadows of various objects are sometimes excessively sharp if only a single light source is used; two or three different light sources with overlapping beams may give better results. In fact, most kinds of merchandise will show up better under a dual system of fluorescent and incandescent lighting than under an all-fluorescent or all-incandescent system.

Show-window lighting is another application where combinations of light sources are desirable. If fluorescent lamps, for example, are used alone, the window display will lack the highlights, shadows, and dramatic contrasts that can be had by using a combination of both incandescent and fluorescent lamps.

The lighting flexibility necessary in stage lighting can only be accomplished by the use of several different light sources. For example, fluorescent lamps may be used to supply a certain level of illumination, but several controllable—in both directional and light intensity—spots and floods are required to achieve the best color combination for dramatizing a setting. Various colors of lamps such as red, blue, green, etc., are also necessary for various effects.

Another important use of light-source combinations is where mercury vapor (MV) lamps are used. As discussed in Chap. 4, even a momentary power interruption will cause the lamps to go out, and it may be 10 to 15 minutes before they will restart and come up to full light output again. Therefore, for safety reasons, any MV lighting installation for building interiors should be supplemented with either incandescent or fluorescent light sources.

In the following pages there is a discussion of several lighting projects where combinations of two or more different light sources were used simultaneously. While these examples barely touch upon the various combinations possible, they will give one an idea of what is possible so that one may approach any such design with greater assurance.

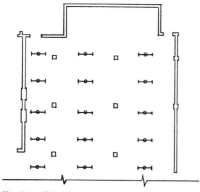

Fig. 5-1 Church basement prior to renovation.

OTTERBEIN METHODIST CHURCH

When the members of the Otterbein Methodist Church of Harrisonburg, Virginia, agreed to renovate their multipurpose basement area, a local architect and his lighting consultant were hired to study the various problems.

There were three clearly defined lighting problems. A general-lighting system was necessary to provide adequate, quality lighting during the men's Sunday School class and similar meetings; a different type or mood of lighting was required when the area was used for church dinners; and finally, indirect lighting was sought to enhance the architectural and decorative features of the area.

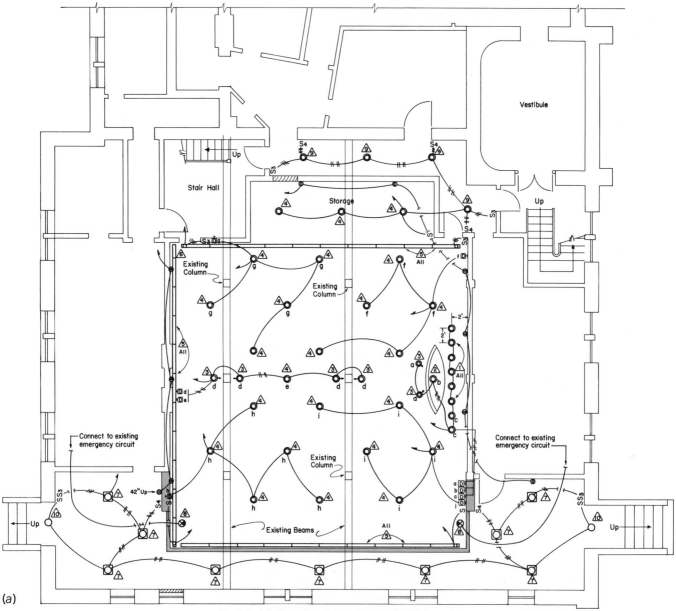

Fig. 5-2 (a) Floor plan of same basement with incandescent recessed fixtures laid out; (b) symbol list; (c) lighting fixture schedule (see page 63).

To these were added stringent qualifications by the church's minister and the building committee. The lighting design not only was to serve the functional needs of the area, but must also be particularly successful in its fusion of the architectural and the spiritual needs.

Figure 5-1 illustrates the problems of the area prior to renovation. The existing surface-mounted two-tube fluorescent lighting fixtures not only gave little or no versatility in lighting control and inadequate illumination, but were also very obtrusive to standing speakers because the ceiling was extremely low for an area of this size. This low ceiling also was a problem to the new lighting design as there was no practical way to raise the ceiling or to lower the existing floor.

An examination of the existing ceiling structure showed that the rough 2- by 10-in. joist was placed approximately 16 in. on center. With this information, the electrical designer laid out a uniform pattern of 300-W incandescent recessed fixtures, as shown on the floor plan in Fig. 5-2a and b. Twenty of these fixtures were controlled (in groups of five) by 2-kW rheostat dimmers. The twenty-first fixture, mounted between the two columns, was controlled separately by a 500-W rheostat dimmer. These 21 incandescent fixtures, having fresnel lenses, and con-

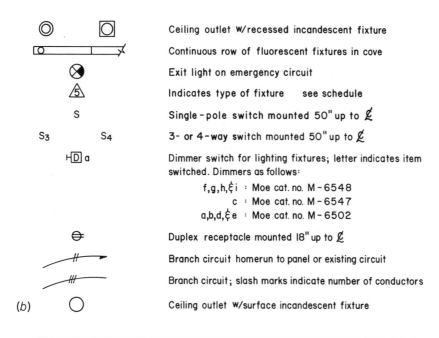

(b)

	Symbol description
◎　　▢	Ceiling outlet w/recessed incandescent fixture
	Continuous row of fluorescent fixtures in cove
⊗	Exit light on emergency circuit
△₅	Indicates type of fixture see schedule
S	Single-pole switch mounted 50" up to ₵
S₃　　S₄	3- or 4-way switch mounted 50" up to ₵

Dimmer switch for lighting fixtures; letter indicates item switched. Dimmers as follows:

f,g,h,&i : Moe cat. no. M-6548
c : Moe cat. no. M-6547
a,b,d,&e : Moe cat. no. M-6502

⊖	Duplex receptacle mounted 18" up to ₵
	Branch circuit homerun to panel or existing circuit
	Branch circuit; slash marks indicate number of conductors
○	Ceiling outlet w/surface incandescent fixture

NOTE: It is imperative that the electrical contractor visit the project site and consult with the architect before the bid date if there are any questions.

Lighting Fixture Schedule						
Fixture Type	Manufacturers description	Lamps		Volts	Mounting	Remarks
		No.	Type			
△1	Halo cat. no. H44-441	1	150W (R-40)	120	Recessed	
△2	Halo cat. no. H7-79	1	100W P-38		Recessed	
△3	Halo cat. no. H7-76	1	150W IF		Recessed	
△4	Halo cat. no. H49-493	1	300W IF		Recessed	
△5	Lithonia cat. no. F-140-RS	1	40W F		Cove	Reflector lamps
△6	Lithonia cat. no. F-120-RS	1	20W F		Cove	Reflector lamps
△7	Halo cat. no. H2-23	1	150W IF		Recessed	
△8	Halo cat. no. H2641-2BZ "G"	2	25W IF		Surface	No direction arrows
△9	Halo cat. no. H56-686	1	150W IF		Recessed	
△10	Moe cat. no. M-879-7	2	60W F IF		Surface	

Fig. 5-3 Lighting fixtures used to highlight paneled wall.

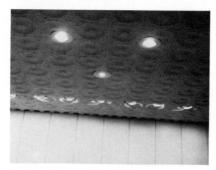

Fig. 5-4 Recessed lighting fixtures used to highlight speaker.

Fig. 5-5 Illustration of lighting-dimmer control.

Fig. 5-6 Recessed lights used to highlight column.

trolled by rheostat dimmers, gave an average calculated foot-candle range of 0 to over 50 fc. This varied range of illumination was adequate, and also provided a good atmosphere for any function that would be held in this area. Since the fresnel lenses were flush with the ceiling, the appearance was not objectionable to either the speakers or the audience. The fixtures blended very well with the accoustical tile ceiling and also maintained a spiritual mood, as can be seen from the accompanying photographs.

The 300-W IF (inside-frosted) incandescent lamps in the 21 recessed lighting fixtures also improved the general appearance of furnishings, people, and especially food. It has been the designer's experience that cool-white fluorescent lamps have always given all food, and especially meats, a very poor appearance.

The various highlights throughout the area were relatively simple to provide. The nine direct wall-wash lighting fixtures (Fig. 5-3) located behind the speaker to highlight the paneled wall behind the speaker come from recessed fixtures; each has a half-ellipsoid reflector, a block Coilex baffle, and a rimless scoop trim. Then, 150-W reflector-type lamps were used to produce a very nice wall-wash effect without glare.

The speaker was further highlighted by two ceiling-recessed fixtures (Fig. 5-4). Each houses a 100-W adjustable narrow-beam reflector lamp. The speaker's podium was lighted by one direct ceiling-recessed fixture with a 2-in. pinhole lens. This lighting fixture provides adequate illumination on the speaker's reading material.

All of the wall-wash and speaker highlight fixtures are also controlled by rheostat dimmers (Fig. 5-5) for added effects and versatility.

The designer had some reservations concerning the highlighting of the speaker with incandescent lamps at such close range. However, while some heat is felt from the lamps, it has been reported that this heat is not too objectionable for a period of 30 to 45 minutes. The dimmer can also be used to control, to some degree, the heat produced by the lamps. There was no objectionable glare at any degree of brilliance with the 100-W lamps, and the highlight effect on the speaker's face and shoulders was more than satisfactory.

The two columns located approximately in the center of the area, as can be seen on the floor plan, were thought to be a great problem to the architect. The columns offered a possible obstruction to the field of vision between the seated audience and the speaker. They were also thought to be unattractive. However, the electrical designer suggested paneling these columns to match the walls, and then highlighting them instead of trying to hide them.

Two ceiling-recessed downlights were mounted on two sides of each column, as shown on the drawings and in Fig. 5-6. This changed what was thought to be a hindrance into an attractive added effect. All four of these lighting fixtures were also controlled by one inexpensive rheostat dimmer.

At this stage, the new lighting design had greatly improved

both the appearance and the illumination levels of the area. Still, the very low ceiling left much to be desired—architecturally speaking.

In an attempt to improve the appearance of this low ceiling, the electrical designer specified a continuous cove along the ceiling line on three sides of the area. A continuous row of 40-W fluorescent strip-lighting fixtures were then specified to be mounted in this cove. Each side was controlled by a single-pole toggle switch. Not only did this addition provide a supplemental light source for still more versatility in desired effects, but these additional cove lighting fixtures gave the low ceiling the appearance of being much higher than it actually was. This, in turn, improved the overall appearance of the area. See Fig. 5-7.

Fig. 5-7 Illustration of cove lighting.

DEPARTMENT STORE

The owner's initial requirements for the electrical system in a recently constructed department store were brief but to the point. "The electrical system shall be designed to provide adequate and uninterrupted power supply for all equipment and devices which require electrical energy. The electrical system shall be simple, economical, and safe."

After attending several preliminary conferences with both the architects and owners present, obtaining all available architectural drawings, and getting a sales fixture layout, the lighting designer began the actual layout of the lighting system.

Modern practice requires the establishment of a minimum quantity of light throughout a given area, which is usually termed "general lighting." This lighting must be arranged so that the eye can function with ease and efficiency. Experienced judgment combined with lighting calculations enabled the designer to create the lighting layout as shown in Figs. 5-8 and 5-9. Notice that the overall lighting arrangement combines fluorescent and incandescent lighting fixtures in order to obtain the desired illumination levels, color quality, and effects. A single light source for this type of sales area would have been entirely inadequate.

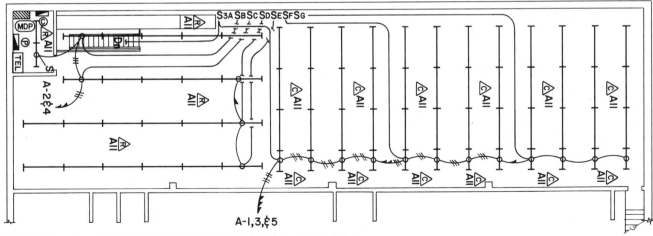

Fig. 5-8 Floor plan of the lighting layout for the mezzanine of a department store.

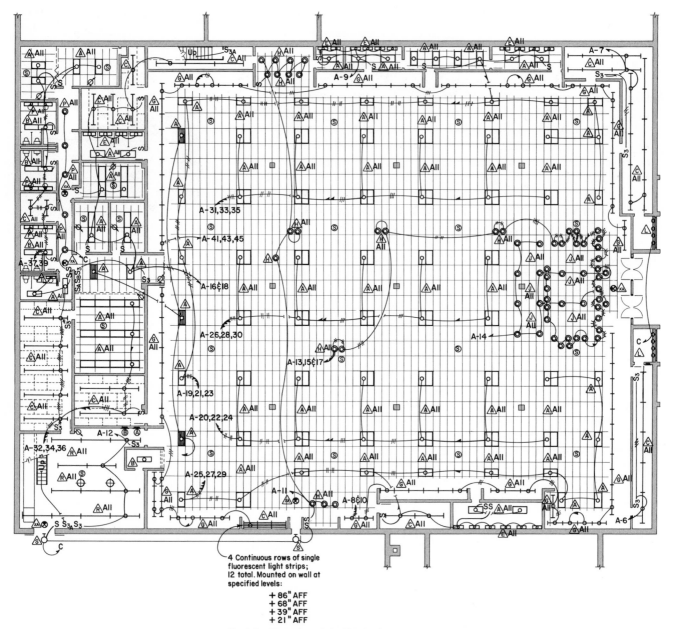

4 Continuous rows of single
fluorescent light strips;
12 total. Mounted on wall at
specified levels:

+ 86" AFF
+ 68" AFF
+ 39" AFF
+ 21" AFF

Fig. 5-9 Floor plan of the lighting layout for a department store.

The paragraphs to follow will explain the reasoning behind the designer's layout for this store building.

Recessed 2- by 4-ft, four-lamp lighting fixtures with acrylic lenses were used throughout the sales area for general lighting. The aim was to obtain a uniform level of illumination of not less than 65 fc throughout the entire area, at 30 in. above the floor.

It is apparent that the layout or location of lighting fixtures largely determines how uniformly the light will be distributed over an area, just as the location of sprinkler heads regulates water coverage in case of fire. The layout illustrated in Fig. 5-10 would have provided more uniform light distribution than the

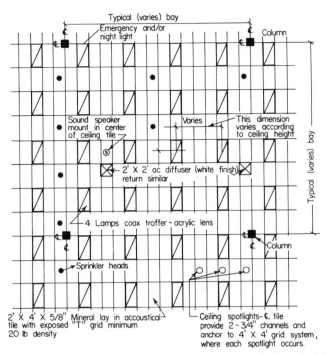

Typical (varies) bay

Emergency and/or
night light

Column

Sound speaker
mount in center
of ceiling tile

Varies

This dimension
varies according
to ceiling height

Typical (varies) bay

2' X 2' ac diffuser (white finish)
return similar

4 Lamps coax troffer - acrylic lens

Column

Sprinkler heads

2' X 4' X 5/8" Mineral lay in accoustical
tile with exposed "T" grid minimum
20 lb density

Ceiling spotlights-℄ tile
provide 2 - 3/4" channels and
anchor to 4' X 4' grid system,
where each spotlight occurs.

Fig. 5-10 A higher-quality lighting layout for Fig. 5-9. (NOTE: All 2- by 4-ft troffers must be supported by 12-gauge galvanized annealed wires from the top and secured to the structure above.)

layout used, but since one of the owner's requirements was economy, the layout actually used was the best compromise.

The design of the general lighting for this store was also governed by the arrangement of bays, columns, beams, and other architectural details; they required that the layout be fitted as symmetrically as possible to the interior but always kept within the limits of spacing directed by ceiling height or the heights at which lighting units could be mounted.

Again, recessed 2- by 4-ft fluorescent lighting fixtures with 100 percent acrylic low-brightness prismatic lenses were used in all office areas. However, in these areas, it was calculated that the maintained illumination level was 150 fc at 30 in. above the floor. The zonal cavity method was used for this calculation.

Lighting in stockrooms and receiving areas was designed to provide an illumination level of 50 fc maintained at 30 in. above the floor. Lighting fixtures in these areas were surface-mounted strips with exposed unshielded fluorescent tubes for economy.

The lighting layout in stairways, as well as in all alteration rooms, was arranged to maintain an illumination level of 100 fc. All lighting fixtures in these areas were acrylic-shielded fluorescents—surface-mounted in some locations and recessed in others.

In quite a few instances, surface-mounted strips with exposed unshielded fluorescent tubes have been used in corridors and stairways, and the lighting result has been unsatisfactory. The reason for this is that the high brightness of this type of lighting fixture causes uncomfortable glare and poor, or no, diffusion.

Two successful methods are generally being used today in lighting corridors and stairways. One is to provide a good matte-white ceiling and to use totally indirect lighting (in some, the light sources are completely hidden); the other is to employ a light source of low brightness which in turn provides for high visual comfort. In stairways, the problem of changing lamps should also be considered when selecting either the type or the location of a lighting fixture.

The show windows near the main store entrance are illuminated to a level of 500 fc by utilizing high-intensity shielded and adjustable spotlighting fixtures. Two or more different light sources in these areas would have provided more versatility for displays, but again economy had to be considered.

All fitting rooms were provided with wall-mounted shielded fluorescent fixtures over the mirror in the dressing booths and a ceiling-mounted acrylic-shielded fluorescent fixture in the center of the fitting area.

Exit lights have stenciled faces and cast-aluminum bodies, and use fluorescent lamps. They are also designed to serve as emergency downlights, providing 5 fc of illumination at the door.

Note that no unshielded fluorescent fixture was installed in any area which would be visited by the public.

All incandescent lamps were either inside-frosted or silver bowl, as required. The fluorescent lamps in the sales area, beauty salon, fitting rooms, toilets, and hall as well as in storage rooms were standard warm white for color quality. Cool-white

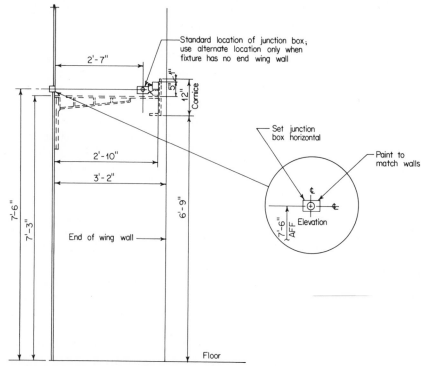

Fig. 5-11 Perimeter-lighting detail (valance lighting) showing alternative "J" box location where no end wall occurs. All equipment shown with broken lines is supplied by fixture contractor. Light fixture is supplied by electrical contractor.

fluorescent lamps were used in all offices, bulk-storage areas, alteration rooms, shipping and marking areas, and stairways. This specification, of course, was to provide the greatest number of lumens per watt.

Supplemental Lighting

The perimeter lighting shown in the floor plans in Figs. 5-8 and 5-9 and also in the detail in Fig. 5-11 served to highlight the merchandise shelves around the perimeter of the store. Although all these lamps were bare tubes, the light sources were not readily visible to the public.

The remaining filament (incandescent) lighting in the sales areas consisted of either recessed downlights or adjustable spots and were used for various merchandise displays as required by the store-fixture layout.

INDOOR VOLLEYBALL COURT

The alterations and additions to the Shenandoah Valley Detention Home in Staunton, Virginia, called for one indoor recreation area, the floor plan of which is illustrated in Fig. 5-12.

Each of the fifteen type "5" lighting fixtures contained a 250-W color-corrected mercury vapor lamp. This type of lighting fixture was also supplied with a lens guard to protect the fixture from damaging impacts caused by a misdirected volleyball.

The eight type "7" lighting fixtures also installed in this area were used for two purposes. One was to supplement the type "5" fixtures for greater lighting intensity; the other was to provide a light source should a power interruption occur, causing the MV fixtures to go out and take 10 to 15 minutes to restart. These type "7" lighting fixtures are surface-mounted, and each contains a 100-W incandescent lamp.

The type "7" fixtures were controlled by one single-pole switch near the entrance to the area, while the three circuits feeding the fifteen type "5" fixtures were controlled by a three-pole 20-A lighting contactor.

DIRT ACCUMULATION

Since this is the last chapter on lighting, the matter of dirt accumulation should be mentioned. No matter how carefully designed a lighting system may be with respect to type and size of lamp and lighting fixture, spacing of fixtures, etc., it must be kept clean if it is to remain in good operating condition.

Dust and dirt allowed to accumulate on reflecting surfaces of lighting equipment may, in the space of a month, decrease the light output from 20 to 30 percent.

The lighting designer can help alleviate the dirt problem by selecting appropriate lighting fixtures for the type of area in question. In addition to cleaning the lighting fixtures frequently, frequent painting and maintenance of the surroundings—walls, ceilings, etc.—will help keep the lighting efficiency near its original value. While most of the suggestions are usually out of the lighting designer's control, the designer can make certain that the owners are aware of the situation.

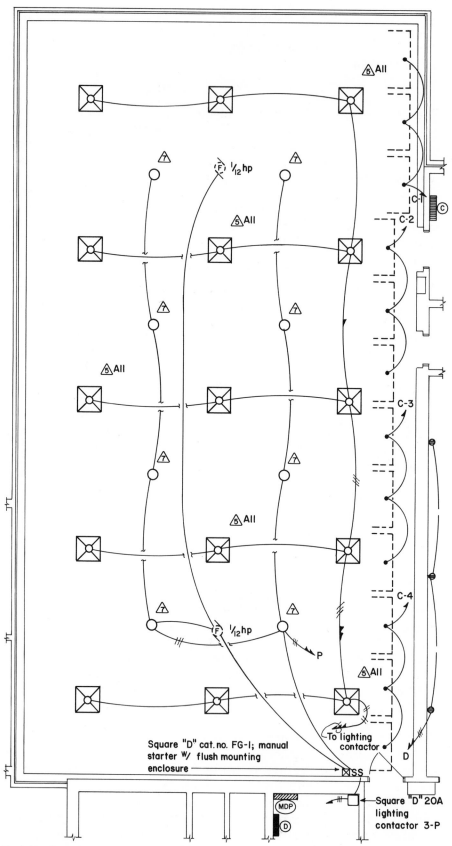

Fig. 5-12 Floor plan of an indoor recreation area.

Chapter 6

Power Wiring for Convenience Outlets

The ordinary duplex receptacle or convenience outlet used in residences, stores, offices, and other buildings is the one electrical device most often used by the occupants. For this reason, it is designed to be one of the safest energy-transmitting devices manufactured. Since the early 1950s, the grounded receptacle has been mandatory on new construction and renovations under the National Electrical Code and was another step forward in further reducing electrical shock hazards.

Article 410-52(d) of the 1971 National Electrical Code states: "All 15- and 20-ampere attachment plugs and connectors shall be so constructed that there are no exposed current-carrying parts except the prongs, blade or pins. The cover for wire terminations shall be mechanically secured, or an integral part of the attachment plug or connector."

As of May 1, 1974, Underwriters' Laboratory is requiring that all plugs not of the dead-front construction type meet this requirement by having the insulating disk secured in place. More stringent requirements can be expected in the future, and they may cover:

1. Wiring contacts
2. Strain relief
3. Impact resistance
4. Gripping configuration
5. Dead-front construction

In a commercial building, the plug receptacles or convenience outlets are usually placed on separate circuits and are not made a part of the general lighting circuit as is sometimes done in a residence. The Code states that a minimum of 180 W shall be allowed for this type of outlet. Therefore, since a duplex receptacle has two plugs, a minimum of 360 W shall be allowed for each duplex receptacle.

Most electrical designers round this figure off and allow 400 W for each duplex receptacle. This makes it easier to keep the figures in one's head while circuiting the outlets and allows a slight margin to ensure sufficient capacity.

The following floor plan (Fig. 6-1) shows a room with several duplex receptacles located at various intervals within the room. Assuming 20-A branch circuits, how many circuits will be required to feed all the receptacles?

First, find the number of receptacles that can be connected to a 20-A branch circuit.

$$\frac{120\ (V) \times 20\ (A)}{400\ (W\ per\ duplex\ receptacle)}$$

$$= 6\ (number\ allowed\ on\ circuit)$$

Since there are twelve duplex receptacles in the area, they will require two branch circuits and may be circuited as shown in Fig. 6-2.

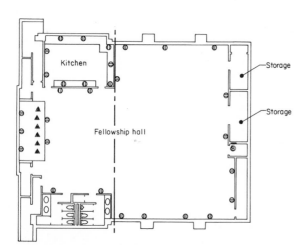

Fig. 6-1 Floor plan showing a room with duplex receptacles located at various intervals.

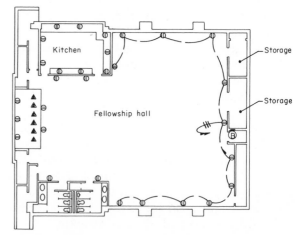

Fig. 6-2 Circuiting of the receptacles shown in Fig. 6-1.

MULTIOUTLET ASSEMBLIES

The National Electrical Code defines a *multioutlet assembly* as a type of surface or flush raceway designed to hold conductors and plug receptacles, assembled in the field or at the factory. "Plugmold" is one of several brand names of multioutlet assemblies available, and is a product of The Wiremold Co.

The Code further states that where fixed multioutlet assemblies are employed, each five feet or fraction thereof of each separate and continuous length shall be considered as one outlet of not less than 180-W capacity, except in locations where a number of appliances are likely to be used simultaneously, when each foot or fraction thereof shall be considered as an outlet of not less than 180 W.

The floor plan shown in Fig. 6-3 gives an example of multioutlet assemblies used in a commercial store area. Multioutlet assemblies were used because several television sets and small appliances were to be demonstrated in these areas. However, multioutlet assemblies are not limited to such locations only, but are also practical anywhere that convenience receptacles are desired fairly close together.

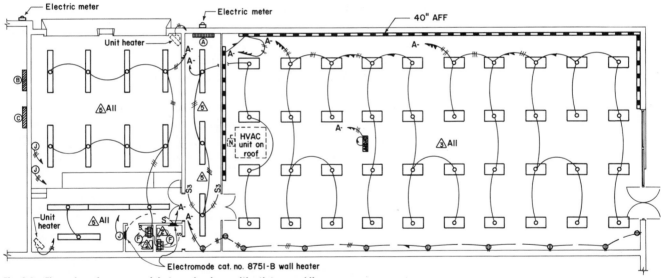

Fig. 6-3 Floor plan of a commercial store showing multioutlet assemblies.

SMALL-APPLIANCE BRANCH CIRCUITS AND RECEPTACLES

The National Electrical Code states that for small-appliance outlets, two or more 20-A branch circuits shall be provided, and "such circuits shall have no other outlets" connected. On these circuits, No. 12 AWG wire must be used to carry the required 20-A load; 1.5 kW is the figure used when calculating residential appliance circuits.

In addition to the two 20-A circuits required for appliances used in the kitchen, pantry, family room, dining room, or break-

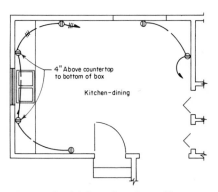

4" Above countertop
to bottom of box

Kitchen–dining

Fig. 6-4 Partial floor plan of a residence.

fast area, the Code requires that "at least one 20-A branch circuit shall be provided for laundry receptacles." This is required because automatic washers draw a large amount of current during certain cycles of their operation. The amount of current drawn from irons is also large, and such appliance circuits will lessen the danger of overloading circuits.

The kitchen floor plan of the residence in Fig. 6-4 shows six duplex receptacles. In accordance with the Code, two circuits would have been sufficient to feed these receptacles, i.e., three outlets on each circuit. However, to provide greater insurance of sufficient capacity for the many small appliances used in the kitchen today, the designer used three 20-A circuits (two outlets on a circuit! to feed these six receptacles.

The receptacle provided for the portable dishwasher was fed by one 20-A circuit because a large amount of current is drawn by this one appliance. The single receptacle for this circuit is also rated at 20 A since the Code states that "a single receptacle installed on an individual branch circuit shall have a rating of not less than the rating of the branch circuit."

The receptacle provided for the garbage disposal was again fed by a single circuit, this time at a 15-A rating. Again, to comply with the Code, a receptacle with a rating of 15 A was used.

Receptacles in the remaining areas of the home are located so that no point along the floor line in any wall space is more than 6 ft from an outlet in that space. Where the sliding glass door extends to the floor, it is not possible to install receptacles in this glass; thus, outlets were installed on each side of the door so that a 6-ft extension cord within the space of the sliding door would reach one of these outlets.

In situations where sliding glass doors or windows extend to the floor and they are more than 12 ft wide, some means other than conventional wall-mounted receptacles must be used to comply with the National Electrical Code. The use of floor outlets is one way to solve this problem.

The bedroom receptacles are split-wired so that the top plug of the receptacle is always energized while all the bottom plugs may be switched on and off by means of a single-pole switch near each door. This enables table lamps to be plugged into the bottom plug and provides a means of controlling them. Such items as clocks, radios, etc., may be connected to the top plug and will remain "hot" all the time.

It is recommended that one convenience outlet be placed in hallways, regardless of their size. In larger hallways, one outlet should be provided for every 15 ft (linear) of hallway. Such outlets are normally used for vacuum cleaners, floor polishers, table lamps, etc.

The National Electrical Code does not specify the maximum number of receptacle outlets to be connected to each circuit in residential applications. However, it is good practice never to load a circuit to more than 80 percent of its current-carrying capacity. In the case of a 15-A circuit at 120 V, this would be $15 \times 120 \times 0.8 = 1.44$ kW. Still it is difficult to predetermine just what will be connected to these outlets in actual use. In residential applications, more designers allow 300 W per duplex receptacle unless it is provided for a known usage. This means four or five outlets per circuit if it is provided with a 15-A over-current protective device. If this figure is used in designing residential electrical systems, the end result will be a well-designed branch circuit with minimal voltage drop.

Other electricians and designers figure $1\frac{1}{2}$ A per duplex receptacle; this method allows a maximum of ten outlets per circuit. Even by this method, common sense indicates that certain outlets will not use $1\frac{1}{2}$ A. These would include outlets for clocks, night-lights, etc.

If such low-wattage items are known to be connected to the circuit, the designer would be justified in increasing the maximum number of ten outlets to possibly twelve outlets per circuit. The reverse is also true. For example, if outlets were to be installed over a home workbench where it is known that heavy power tools will be used, the designer will most certainly provide outlets and circuits to handle this load. The outlets and circuits will probably be similar to those installed in the kitchen or laundry area.

The only problem with allowing for small wattage loads such as night-lights and clocks is that the load may change in the future. The night-light may be removed from a bedroom outlet, and a window air conditioner may be connected. This is why most designers like to play it safe and allow 300 W minimum for each duplex receptacle.

In laying out receptacle outlets for residential applications, the first step is to provide outlets for known usages such as washer, dryer, dishwasher, workbench outlets, bedside outlets, etc. The information can usually be obtained from the architect, owners, and /or interior decorator. Once the outlets are located for the known requirements, other outlets should be located so that no point along the floor line in any wall space is more than 6 ft from an outlet in that space.

Circuiting of the outlets should be such that all home runs are as short as possible, and all looping between outlets should also be made as short as possible. But on any given electrical installation, there are many different combinations or groupings of outlets into circuits—all of which will be technically correct; the problem is to determine quickly the route which will conserve the most wire and in turn keep voltage drop to a minimum.

The floor plan of a small branch bank is shown in Fig. 6-5. This drawing is titled "Main Floor Plan—Power" to indicate that all power wiring is shown on this plan while the plan showing the lighting layout will be titled "Main Floor Plan—Lighting." On all drawings of electrical systems, except those of warehouses and similar projects where very little electrical work will be required, it is best to use separate floor plans for power and lighting. This enables the drawing to be more easily read by the workers installing the electrical system.

Notice that the designer has laid out the majority of the duplex receptacles in this branch bank in groups of four; that is, the four receptacles in each office are fed by one circuit, and

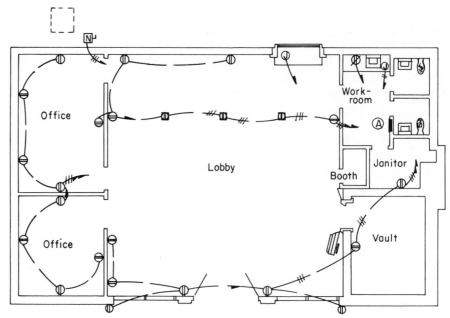

Fig. 6-5 Floor plan of a small branch bank: Main Floor Plan—Power.

then both circuits are run to the panelboard "A" in one conduit using a common neutral.

All the duplex receptacles in the lobby area are also run in groups of four, picking up the two outside outlets at the front entrance in the process. All these circuits are again run to the panelboard, using a common neutral for each two circuits. In a three-phase, four-wire, wye-connected service, most designers try to make all home runs consist of three circuits and a common neutral, but this particular bank is using a three-phase, four-wire, delta-connected service where the high leg cannot be used for 120-V loads.

The duplex receptacle over the countertop in the workroom is indicated by the standard receptacle symbol with a slash through it. This outlet is connected to a separate circuit because the bank personnel will use it for a 1,000-W coffeemaker. The junction box adjacent to this outlet feeds a small electric water heater under the counter.

The junction box shown in the cutout section off the lobby area feeds a prefabricated drive-in unit, while the nonfusible disconnect with a three-wire circuit, located outside the building, is provided for a condensing unit which is indicated by a broken-line square.

In general, most electrical designers lay out convenience outlets for commercial applications in much the same manner as for residential; that is, they locate outlets for known needs first, then provide additional outlets so that no point along any usable floor line is more than 6 ft from any outlet. Although this is not a Code requirement, it is good practice.

Common sense, however, will normally dictate the number and location of outlets in most commercial applications. Common sense and good judgment will come with experience, but in the meantime, the new designer should look at existing installations, ask questions concerning these existing outlets, and then attempt an original design.

Figure 6-6 shows another floor plan: a laundry building used at a nursing home. Again, the drawing is labeled "Main Floor Plan—Power" and is drawn to a scale of ⅛ in. = 1 ft 0 in.

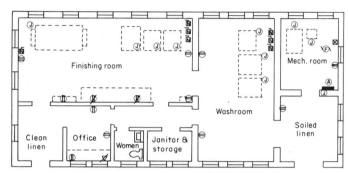

Fig. 6-6   Floor plan of a laundry building: Main Floor Plan—Power.

Each of the junction boxes—indicated by a circle with the letter J inside—contains separate feeders for various pieces of equipment with known loads. The duplex receptacle in the mechanical room is for maintenance purposes, that is, to connect drop cords, hand power tools, etc. The one duplex receptacle in the soiled-linen room and the two receptacles in the washroom were provided for convenience. They will probably be used for vacuum cleaners or similar loads, and therefore all three will be connected to one circuit.

If one goes to the finishing room from the washroom, the duplex receptacle on the left immediately beyond the entrance of the finishing room is provided for a water cooler. The receptacle on the right beyond the entrance, as well as that on the opposite wall, is for convenience. The two outlets over the workbench are for electric clothes- or linen-mending tools, and the one next to the bench is for a sewing machine.

The three duplex receptacles in the office are fed by one separate circuit. The symbol consisting of an open triangle with a slash indicates a telephone outlet above the countertop.

A floor plan titled "Main Banking Level—Power" is shown in Fig. 6-7 and is drawn to a scale of ⅛ in. = 1 ft 0 in.

On this drawing, conventional wall-mounted duplex receptacles are indicated by the usual symbol. The floor receptacles, however, are indicated by a circle with a dot inside. Floor-mounted telephone outlets are indicated by a circle enclosing a shaded triangle.

If one enters the vestibule from the outside, there is one duplex receptacle in the small lobby to the right to provide power for a water cooler; it is fed by one circuit—L-19. The "L-19" indicates that the home run goes to panel L (shown in the opposite vestibule) and that this particular circuit is connected to circuit 19 within the panelboard.

The two floor outlets in the vault-officer area, as well as the single floor outlet mounted under the lobby desk, are connected

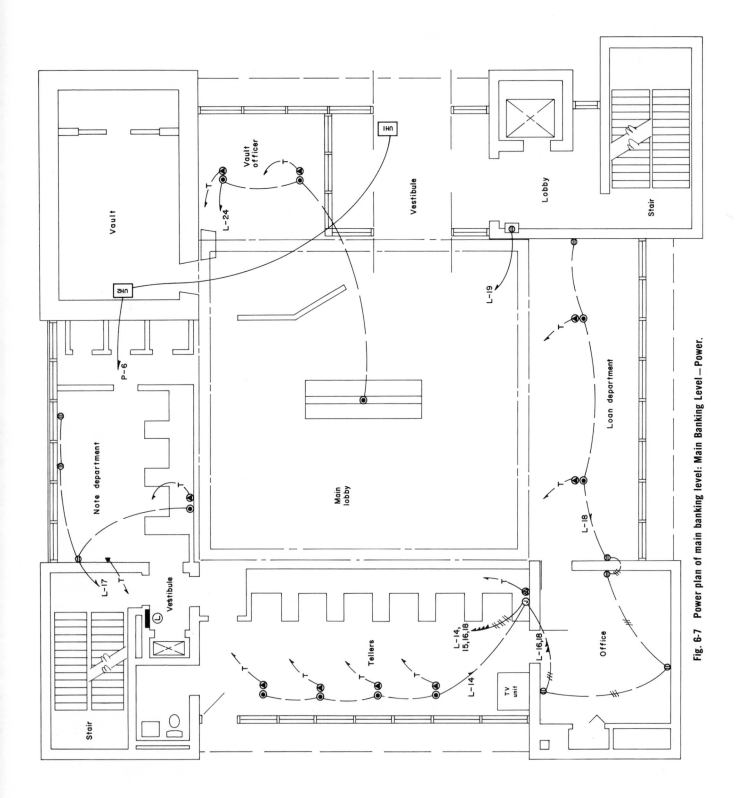

Fig. 6-7   Power plan of main banking level: Main Banking Level—Power.

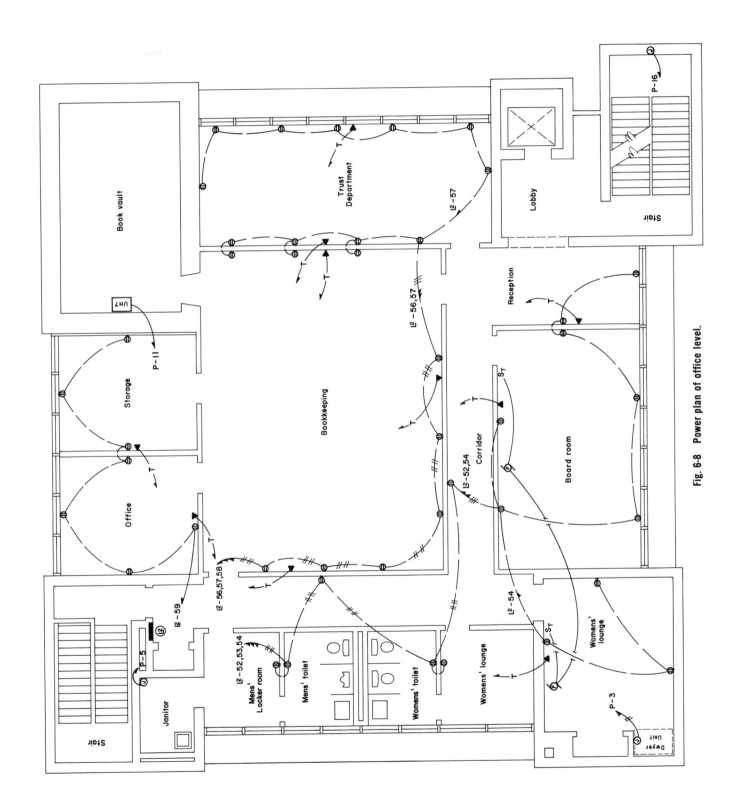

Fig. 6-8 Power plan of office level.

to one circuit that also runs to panel L, but this time it is connected to circuit 24.

All other floor- and wall-mounted receptacles are connected to circuits and run to panel L. The two wall-mounted duplex receptacles and the three floor receptacles in the loading department area, as well as the receptacles in the adjacent office and teller area, run to a junction box before a home run is run to panel L. Since there are four circuits in this home run, six conductors must be provided, because on a three-phase, 4-wire wye-connected service only three circuits may be installed on a common neutral (this would consist of four wires—three hot wires and a neutral). If another circuit is pulled in the same raceway, such as in our case, another hot wire and another neutral are required. Thus, six conductors are required, as indicated by the six slash marks in the circuit on the drawings.

The circuit designated P-6 in Fig. 6-7 feeds the blower motors in unit heater 1 (UH1) and unit heater 2 (UH2). Each telephone circuit home run consists of a $3/4$-in. empty conduit with a galvanized pull wire in each; this may be designated either in the symbol list or in the written specifications.

Three wall-mounted duplex receptacles and one floor-mounted receptacle in the note department are all fed on one circuit running to panel L, as shown, and connected to circuit 17.

Although the designer did not specify an outlet in the stairs or vestibule, one would be advisable for use by the janitor for connecting floor polishers, vacuum cleaners, and the like. Also an outlet near the sink in the lavatory may be useful for those who may want to use an electric shaver.

Another level of the bank just described is shown in Fig. 6-8. On this level (designated Office Level) all duplex receptacles and telephone outlets are wall-mounted. Since most of the outlets were provided for low-power-consuming electronic calculators, up to seven duplex receptacles are shown on a circuit.

The telephone outlets are again fed by an empty $3/4$-in. conduit with a galvanized pull wire. The pull wire is to enable the telephone company personnel to more easily "fish" their telephone cables through the empty conduit.

All circuits on this level run to panelboard L2, and are numbered according to the designated circuits on the panel.

The junction box with a separate two-wire circuit (208 V) is for a combination sink-refrigerator-cooking unit. Another circuit is required to feed the two exhaust fans shown on the drawing, one in the board room and the other in the women's lounge.

Chapter 7
Power Wiring for Special Outlets

The electric motor provides a means of converting electrical energy into mechanical energy and is classified according to:

1. Size, in horsepower
2. Type of application
3. Electrical characteristics
4. Mechanical protection
5. Type of cooling
6. Speed and speed control

Electric motors range in size from small, fractional-horsepower 120-V motors to very large, high-voltage synchronous motors. They are designed for light or heavy duty and also for general-purpose, definite-purpose, or special-purpose types of application. Electrical characteristics will include voltage, frequency, alternating or direct current, type of wiring, and phase (in the case of ac motors). Mechanical protection and type of cooling usually refer to the motor housing, that is, open machine, protected machine, dripproof machine, splashproof machine, totally enclosed machine, explosionproof totally enclosed fan-cooled machine, etc.

Starting and speed-control equipment varies from simple, thermal tumbler switches to large, switch-gear-type controllers. The type of controller depends basically upon the motor to be controlled, the type of control circuit, and the location of the controller. Usually the controller is mounted adjacent to the motor or in its immediate vicinity. Sometimes, however, the controller may be located some distance away, e.g., in a motor-control center. The control center may consist of a group of individual motor controllers or a factory-built self-contained unit having all the motor controllers installed in it.

Motor controllers are available for manual or automatic operation, either at the controller or at a point remote from the controller and the motor. Any number of remote push-button stations may be used with automatic controllers.

When a disconnecting means is required for the motor, it may be a separate externally operated switch, or it may be combined in the same enclosure as the controller. In some instances the motor circuit disconnecting means may serve as the motor disconnect.

The garbage disposal shown by symbol and notation on the floor plan in Fig. 7-1 is rated at 7.5 A at 120 V. A separate 15-A, 120-V branch circuit is shown going to panel A from the junction box provided for connection purposes on the disposal unit. Most garbage-disposal units are powered by split-phase, 120-V motors of between 1/3- and 1/2-hp rating; and according to the National Electrical Code, running-overcurrent protection is required and must not exceed 125 percent of the full-load current rating of the motor.

The garbage-disposal unit shown in Fig. 7-1 was provided with a running-overcurrent protection by using a thermal protector ("heater") built into the unit. This thermal protector is of the manual-reset type. In other units, however, automatic resetting of the thermal protector is featured.

Where running-overcurrent protection is an integral part of the disposal unit, the electrical designer merely shows a circuit feeding the unit, and a regular toggle switch if needed for control. However, where running-overcurrent protection is not an integral part of the disposal unit, the designer must specify one such as that shown on the drawing in Fig. 7-2.

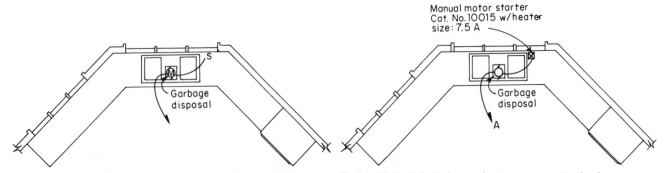

Fig. 7-1 Floor plan showing garbage disposal by symbol and notation.

Fig. 7-2 Method of showing running-overcurrent protection for a garbage-disposal unit on working drawings.

EXHAUST FANS

Exhaust fans are another type of motor-driven device that uses electric current for power. They may be equipped with an integral pull-chain switch for starting and stopping, or a separate wall switch may be used for control. When an automatic system of humidity control is desired, these fans may be controlled with a humidistat.

Single-speed or multispeed exhaust fans used for ventilating residential kitchens, bathrooms, utility rooms, or laundry rooms have a very small power demand and are usually connected to lighting circuits. Larger fans for commercial kitchens or industrial applications, however, require motors of several horsepower and usually are controlled by an elaborate motor control system.

The floor plan of the "Parking Level" of a bank building is shown in Fig. 7-3. The area labeled "Vest.," or vestibule, shows an exhaust fan (F) connected to panel P and controlled by a safety switch on a wall of this area. Note that this circuit is connected to circuit 10 of panel P since the home run is designated "P-10."

Power outlets in Fig. 7-3, other than duplex receptacles, include circuits feeding the blower fans for unit heaters UH3, UH4,

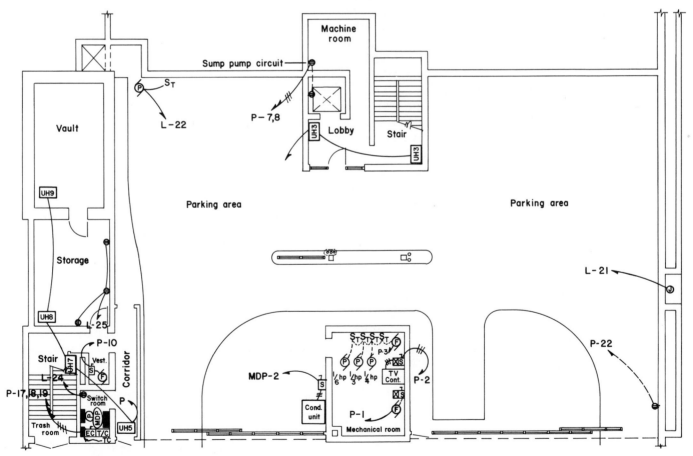

Fig. 7-3 Floor plan of the parking level of a bank building.

UH5, UH7, UH8, and UH9; a junction box connected to circuit L-21 for a telephone booth; a three-phase, 208-V, three-wire circuit (MDP-2) for a pad-mounted condensing unit; two other exhaust fans in the mechanical room (P-1 and P-2); a three-phase, 208-V, three-wire circuit for the closed-circuit television controls; a 120-V circuit feeding three small circulating pumps in the mechanical room; and another circulating pump (L-22) in another area.

All the circuits in Fig. 7-3, shown with solid lines, are to be installed concealed in the ceiling; those designated with a short dashed line are run in the ceiling or on the wall concealed, while those symbolized with the long dashed lines—such as shown for the receptacles in the storage area—are run concealed in the floor or slab.

A time clock used to control certain lights in the parking area is fed with three circuits connected to panel P, circuits 17, 18, and 19.

WATER HEATERS

Many water heaters used for residential and commercial applications require electric current for operation. The residential size may range from 1.2 to 4.5 kW, the smaller size at 120 V and all others at 240 V. Today the average residential electric water heater consists of a 52-gal tank with a maximum wattage of 4.5 kW. Depending upon the length of the circuit, the minimum wire size should be No. 10 AWG.

Water is heated in the storage tank by means of copper-clad electric heating elements which enter the side of the tank. These elements (usually two) project inward and are immersed in the water.

When two elements of 4.5 kW each are used in an electric water heater, the controls are usually wired so that only one element will operate at a time. That is, when the bottom thermostat calls for heat, the contacts close and allow current to flow through the heating element. Whether or not the upper thermostat is calling for heat, the contacts will not close as long as current flows in the lower element.

When the desired temperature is reached at the bottom thermostat, the contacts open and stop the flow of current. At this time, if the upper thermostat still calls for heat, the upper contacts will close and allow current to flow through this element. This method of control prevents the existence of a load of more than 4.5 kW. A wiring diagram showing the connections of the circuit just described is shown in Fig. 7-4.

In some cases it may be desirable to permit both the upper and the lower elements to become energized simultaneously should the thermostats indicate the need. This is true where a quick hot-water recovery is desired. In this case, two circuits capable of carrying the total amperage of each element should be provided; that is, each element will have its own circuit.

Each storage tank of a large commercial water heater may contain six or more elements with a rating of 20 kW or more. In such cases, the manufacturer's wiring diagram of the specific unit should be referred to before the electrical feeders are de-

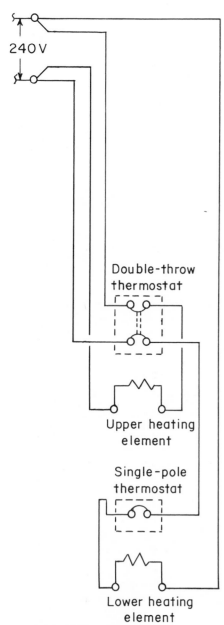

Fig. 7-4 Wiring diagram showing the connections for a water heater.

signed. The unit may require one large feeder, or it may require several smaller circuits for each element. One such wiring diagram is shown in Fig. 7-5.

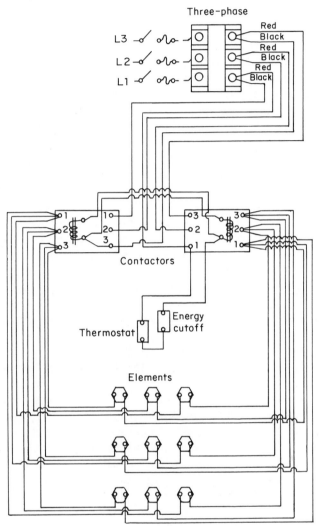

Fig. 7-5 Wiring diagram of a large commercial water heater.

CLOTHES-DRYER CIRCUITS

In addition to the receptacle circuits described in Chap. 6, most residences need circuits for special appliances of which the clothes dryer is one. The clothes-dryer circuit usually will be a 120/240-V, single-phase, three-wire circuit of No. 10 AWG wire, since most of the heating elements are 4.5 kW. A clothes-dryer circuit is shown in the residential floor plan in Fig. 7-6.

The clothes-dryer outlet in Fig. 7-6 indicates a 30-A, 250-V dryer receptacle. This is the most common method of providing electric power to clothes dryers. The dryer is then connected by plugging in a cord attached to the dryer.

Another method is to indicate a three-wire circuit run to the dryer location and then connected directly to a junction box on

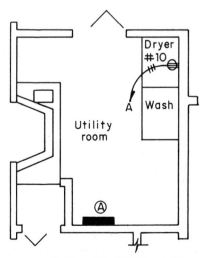

Fig. 7-6 Method of showing clothes-dryer circuit on a residential floor plan.

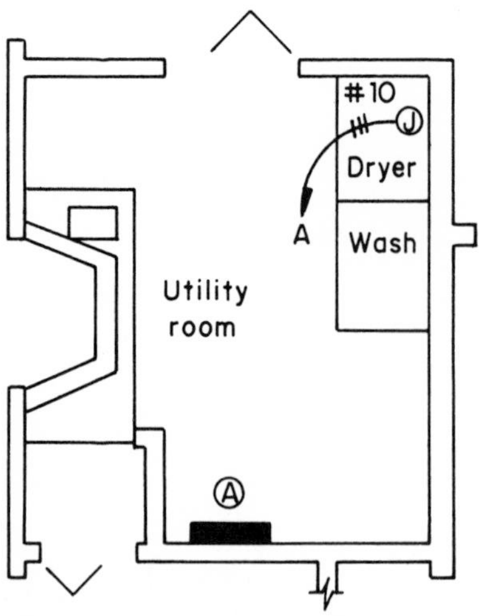

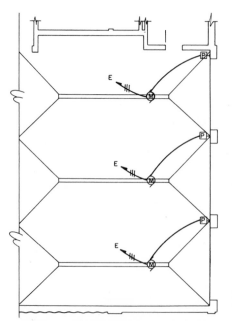

Fig. 7-7 Another method of indicating a three-wire clothes-dryer circuit.

the dryer provided for that purpose. The plan view in Fig. 7-7 shows how this is indicated on a working drawing.

OVERHEAD GARAGE-DOOR OPENER

The motor symbols shown in the floor plan in Fig. 7-8 represent motors to operate the overhead garage door openers of this fire station.

Notice that a push button is located at each door to operate its respective door; that is, the three-button switch raises, stops, or lowers the door. Other push-button stations are also located in the console in the dispatch room and are connected in parallel, with the one at each door, so that each door can be controlled from either location. An override switch is also provided in the console so that each group of doors may be raised simultaneously in case of an emergency. The wiring diagram for the control of these doors is shown in Fig. 7-9.

COOKING TOPS AND OVENS

The National Electrical Code permits both the cooking top and the wall-mounted oven to be connected to one circuit provided the circuit has ample capacity to serve both appliances. The diagram in Fig. 7-10 shows how this might be applied to a built-in cooking top and oven in a residence.

The No. 10 AWG tap conductors from the No. 6 AWG conductors must be as short as possible, that is, "no longer than necessary for servicing." Therefore two appliances of approximately 30-A rating each may be fed by one 50-A circuit.

However, since tapping or splicing the smaller-size conductors from the 50-A circuit will require extra junction boxes, cable or conduit connectors, and wire connectors, as well as additional labor for making the splices and connections, it is our opinion that it would be best to specify two separate 30-A circuits if the distance to the panelboard were of reasonable length. The two separate circuits would also offer greater protection in the event of a ground fault because of the lower-rated individual overcurrent protection.

ELECTRIC RANGES

Single electric ranges are wired in much the manner of clothes dryers in that a single circuit (usually rated at 40–50 A) is run from the panel to a 250-V, 50-A "range receptacle." The electric range is then connected by means of a range cord, although the circuit can be connected directly to a junction box on the range provided for such a purpose. Figure 7-11 shows how a direct-connection circuit may be indicated on a working drawing, while Fig. 7-12 shows how the outlet would appear should a range receptacle be desired.

COMMERCIAL-KITCHEN POWER OUTLETS

Most commercial kitchens require a large number of electrical outlets to operate the various pieces of kitchen equipment, and because these outlets are special, a special way of showing the information on the electrical drawings is required.

Fig. 7-8 Floor plan showing location of motors to operate overhead garage doors.

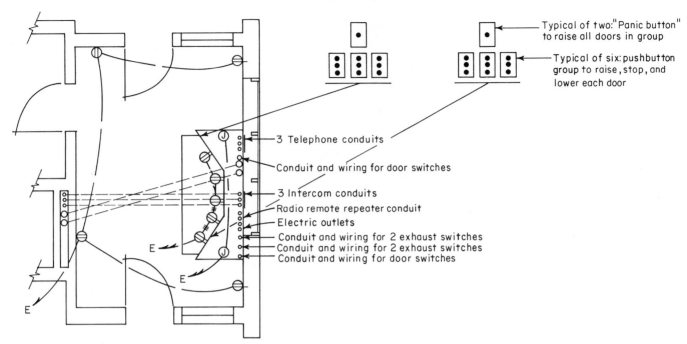

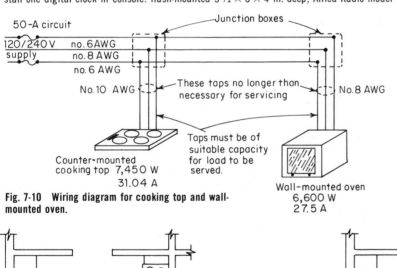

Fig. 7-9 Wiring diagram for the controls of the motors shown in Fig. 7-8. (NOTE: All wiring in console shall be run beneath floor. Use chase in wall directly behind console to run upward; provide and install one digital clock in console: flush-mounted 3½ × 6 × 4 in. deep; Allied Radio model ED-728-6.)

Fig. 7-10 Wiring diagram for cooking top and wall-mounted oven.

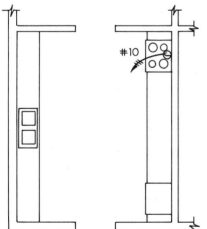

Fig. 7-11 Method of showing direct-connection circuit on a working drawing.

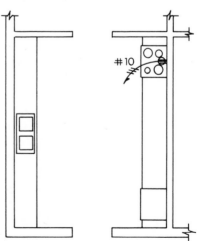

Fig. 7-12 Method of showing an outlet connection on a working drawing.

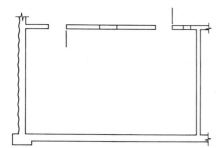

Fig. 7-13 Floor plan of a firehouse kitchen.

A floor plan of a kitchen is shown in Fig. 7-13. This particular kitchen is that of a firehouse and is used in relation to the adjacent meeting room for dinners and dances.

Because of the large number of outlets required in the kitchen in Fig. 7-13, the original drawing is too small to show all the necessary details of construction without becoming confusing. Therefore, the designer had the draftsman enlarge the floor plan to a scale of ¼ in. = 1 ft 0 in.; this is shown in Fig. 7-14.

After the kitchen floor plan was drawn to a larger scale and the pieces of kitchen equipment were drawn, each piece of equipment was given an identification number enclosed in a hexagon, as shown in Fig. 7-14.

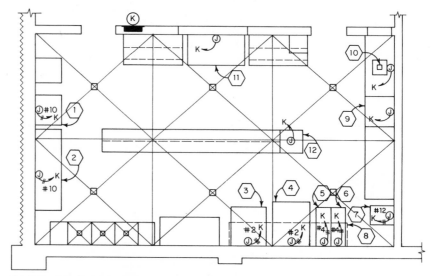

Fig. 7-14 Detailed floor plan of a firehouse kitchen.

Those pieces of kitchen equipment requiring electric power were provided with the symbol to indicate that a junction box or a direct connection is required. The written specifications accompanying the drawings instructed the contractor to verify exact requirements with the kitchen-equipment supplier, that is, the mounting height of the outlets, etc.

From each junction box, a home run is indicated to panel K. The absence of slash marks indicates a circuit of two No. 12 AWG conductors. A circuit other than this is indicated by slash marks representing the number of conductors in the circuit, and the size of the conductors is indicated by a numeral directly adjacent to the slash marks.

The identifying numbers, the junction boxes with home runs, and panel K are also shown in the floor-plan view of the kitchen in Fig. 7-14. The drawing, however, still lacks some of the information required for the electrical installation. First, the contractor needs to know the name of each piece of equipment, the electrical load required for each, and the overcurrent protection required for each.

The kitchen equipment schedule shown in Fig. 7-15 is one method of conveying this information to the contractor.

Kitchen Equipment Schedule								
Equip. no.	Designation	kW or Hp	Volts	Connection			Furnished by	Remarks
				Wire	Conduit	Prot.		
1	Freezer	2 kW	120	#10	3/4"	30A	Others	Provide outlet
2	Refrigerator	1.7 kW	120	#12	3/4"	20A		Provide outlet
3	Range	32 kW	208	# 2	1/2"	100A		Direct connection
4	Oven	18 kW	208	# 2	1 1/2"	100A		Direct connection
5	Fryer # 1	22 kW	208	# 4	1"	70A		Direct connection
6	Fryer # 2	22 kW	208	# 4	1"	70A		Direct connection
7	Mixer	1 Hp	208	#12	3/4"	20A		Provide outlet
8	Kitchen hood	1 Hp	120	#12	3/4"	20A		Direct connection
9	Dishwasher	5 kW	208	#12	3/4"	20A		Direct connection
10	Disposal	1 Hp	208	#12	3/4"	20A		Provide outlet
11	Hot-food table		208	#10	3/4"	30A		Provide outlet
12	Ice-maker			#10	3/4"	30A		Provide outlet

Fig. 7-15 Kitchen equipment schedule for Fig. 7-14.

The "Equip. no." column lists the identifying numbers of each piece of equipment shown in the floor-plan view in Fig. 7-14. The second column, "Designation," names the corresponding pieces of equipment, while the next five columns indicate the horsepower or kilowatts, volts, wire size, conduit size, and over-current protection, respectively. The next column indicates that all the pieces of equipment are furnished by others. The final column indicates the type of connection; that is, direct connection or outlet.

Therefore, the floor-plan view in Fig. 7-14 and the kitchen equipment schedule in Fig. 7-15, combined with the written specifications, leave little doubt as to the exact requirements for the electrical installation for the area. A panelboard schedule, shown elsewhere on the drawings, also indicates on which circuit each piece of equipment is connected, the trip amperes of each circuit breaker, and the type of frame required.

SCHEDULES

Since we have just described one type of kitchen equipment schedule, a word about schedules in general is warranted at this point.

A schedule is a systematic method of presenting notes or lists of equipment on a drawing in tabular form, and when properly organized, it is a powerful time-saving device for the draftsman and designer. A good schedule will also save valuable time for the specification writer as well as the worker on the job. For example, the kitchen equipment schedule shown in Fig. 7-15 lists all the information necessary for a correct installation.

Sometimes all the same information can be found in the written specifications of a project, but the workers do not always carry the specifications, whereas they usually do have constant access to the working drawings. Therefore, the schedule is an excellent means of providing essential information in a clear, concise, and accurate manner, allowing the workers to carry out their assignments in the least amount of time.

Other types of electrical schedules can be found in other chapters and in Appendix B of this book.

Chapter 8

Power Wiring for Heating, Ventilating, and Air Conditioning

Circuit requirements for HVAC (heating, ventilating, and air conditioning) equipment do not greatly differ from those for any other circuit or feeder, but the electrical designer must carefully study the manufacturer's wiring diagram for the equipment in question.

For example, an electric furnace may have a total rating of 24 kW. It would then be natural for the designer to calculate the circuit size as follows for a 208-V three-phase system:

$$\frac{24,000 \text{ W}}{\sqrt{3 \times 208 \text{ V}}} \times 1.25 = 83.33 \text{ A}$$

Using this as a basis, the designer would specify a conductor rated at 83.33 A minimum. However, the resistance-type heaters in a furnace of this type will usually be connected in steps, that is, four electric heaters of 6 kW each. If one conductor is used for the feeder, another subpanel would have to be installed in order to separately feed each of the four heaters. If the designer did not specify this, the electrical contractor would be entitled to an extra billing and would cost someone additional money.

Of course, the electric furnace may have its own built-in controls dividing the 83.3-A circuit into four subcircuits, but the information must be verified with the specifications furnished by the equipment manufacturer before the electrical design is made.

If the electric furnace under discussion required that the circuit be divided into four 30-A circuits, for example, the designer could specify a 100-A subpanel near the furnace, and then use three three-pole 30-A circuit breakers to feed the four electric resistance heaters. It may be best, in some cases, to run the four separate circuits from the original panel directly to the contactors controlling the heating elements. But, the point we are trying to make is that all circuits for HVAC equipment must be verified in order to ascertain the exact requirements.

RESIDENTIAL ELECTRIC HEATING SYSTEM

Residential electric heating units are available in baseboard, wall, floor-mounted, and ceiling heat cable—to name a few—and all are becoming very popular as a means of heating residential buildings. Some of the advantages of electric heat follow.

1. Electric heat is safer than combustible fuels.
2. It is quiet in operation.
3. The initial installation cost is usually lower since electric heat requires no chimneys, little if any storage space, and no fuel tanks.
4. It is the cleanest type of heat available.
5. Each room may have its own individual thermostat which enables the rooms to have variable temperatures to suit each occupant.

Electric heat may also be supplied by a duct system similar in every respect to a conventional oil, gas, or cool hot-air system except that the heat comes either from electric heating elements or a heat pump. When a ducted electric heating system is used, humidity control, air conditioning, and ventilation may also be combined with the system.

In studying the floor plan in Fig. 8-1, we find that there are three ventilating units in the bay area of a firehouse which must have power supplied. These units are actually installed on the roof above the bay area. Each unit consists of a steel housing designed to be mounted on a flat roof and a blower unit which is enclosed in the steel housing.

The blower unit is made up of a single-phase 208-V motor driving a propeller-type fan by means of a V-belt drive. This arrangement results in quieter operation than using a direct drive. A section of this fan, showing the internal construction as well as the electrical connecting box, can be seen in Fig. 8-2. The circuit is shown in the floor plan in Fig. 8-1 and is drawn to indicate a home run to panel P. By referring to the panelboard schedule for panel P, we can determine the type

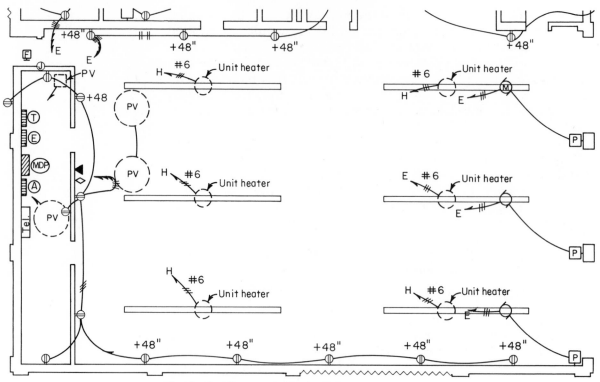

Fig. 8-1   First-floor power plan of a firehouse.

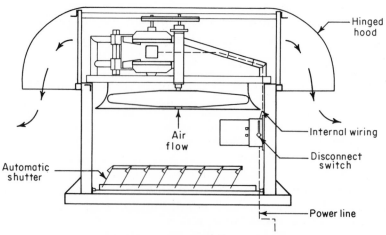

Fig. 8-2   Section of a ventilating fan showing electrical connections.

of circuit breaker provided for overcurrent protection; in this case, it is a two-pole, 20-A circuit breaker. The conductors consist of two No. 12 AWG wires in ½-in. conduit.

The power-riser diagram for this firehouse is shown in Fig. 8-3. Panel H and panel P—one located in the mechanical room and the other located on the mezzanine level—furnish an adequate power supply for all heating and cooling equipment such as electric unit heaters, condensing units, air-handling units, etc., all of which are shown on the floor plan in Fig. 8-1.

The unit heaters are three-phase, 208-V and are rated at 15 kW each. Therefore a 1-in. conduit containing three No. 6

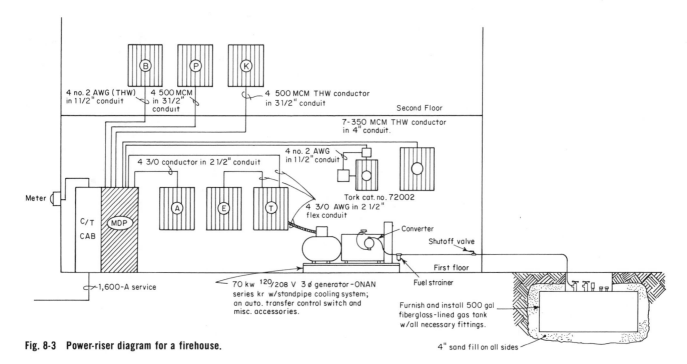

Fig. 8-3 **Power-riser diagram for a firehouse.**

AWG THW conductors was needed to feed each heater. As mentioned previously, all the unit heaters obtain their power from panel H with the exception of two. These two heaters are run to panel E, which is connected to the 70-kW emergency generator in case of a power failure. In the event that a power failure occurs, the generator controls will automatically start the generator and at the same time transfer the loads on panel E to the generator. While the two unit heaters on the emergency panel will not provide comfortable temperatures in the bay areas during extremely cold days, they will keep the temperature above freezing, which is very important in a building such as the one described.

Since most heating and cooling equipment is installed by a "mechanical" or heating and cooling contractor, the electrical plans only need to show the feeder circuits. Even the control wiring of the various pieces of heating and cooling equipment is usually the responsibility of the heating and cooling contractor and are therefore not shown on the electrical drawings.

Motor starters are items which have been confusing for both electrical contractors and heating and cooling contractors, in that the starters are usually furnished by the heating and cooling contractor, but are usually mounted and connected by the electrical contractor. The reasons are that the mechanical contractor should be responsible for obtaining the correct starter for the piece of equipment furnished; yet the mounting and wiring of the starter should be the electrical contractor's responsibility since experienced personnel should work on the voltages usually encountered in motor starter wiring—usually 240 to 480 V.

At times an electrical contractor will be responsible for the entire installation of an electric heating system such as the one shown in the floor plans of a church-education building in Figs. 8-4 and 8-5. The reason is, of course, that no ductwork is involved in this installation, and the job is entirely related to the electrical industry.

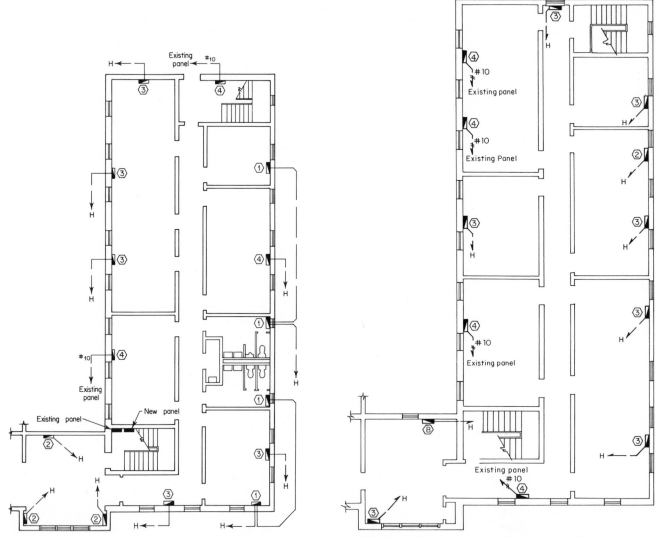

Fig. 8-4 First-floor plan of a church-education building.

Fig. 8-5 Second-floor plan of a church-education building.

The first-floor plan in Fig. 8-4 shows several wall-mounted forced-air heaters indicated by the symbol ⬡. An identifying number is also located at each heater. By referring to the electric heat schedule in Fig. 8-6, we see that the manufacturer, catalog number, dimensions, voltage, mounting, and wattage of each type can be found.

This is an existing building; therefore, to save labor, all wiring will be run exposed, with cutting and patching of existing structures.

To conceal as much of the wiring as possible, it was decided to run the circuits for the heaters on the first level in a trench

Electric Heat Schedule					
Htr. type	Manufacturer's description	Dimensions	Volts	Mounting	Remarks
①	Electromode cat. no. 5990-T mod. no. WA-15	14 9/16" X 11 1/4" X 3 3/4"	240	Surface	1,500 Watts
②	Electromode cat. no. 5990-M mod. no. WA-20	14 9/16" X 11 1/4" X 3 3/4"			2,000 Watts
③	Electromode cat. no. 5990-N mod. no. WA-30	18 1/4" X 14 1/4" X 3 7/8"			3,000 Watts
④	Electromode cat. no. 5990-W mod. no. WA-40	18 1/4" X 14 1/4" X 3 7/8"	↓	↓	4,000 Watts

Fig. 8-6 Electric heat schedule.

outside of the building and to feed the cable through the wall at each heater. A note on the drawing reads as follows:

NOTE: All electric wall heaters on the first-floor level shall be fed with type U. F. cable run in a shallow trench outside of building. See detail on this sheet for connection.

The detail drawing appeared as shown in Fig. 8-7.

The second-floor plan shown in Fig. 8-5 required a slightly different method of feeding, since there was no place to conceal the cable because it was on the first floor. A note appeared on the drawing as follows:

NOTE: Electric wall heaters on second-floor level shall be fed by one 1¼-in. conduit from panel H up through corner of stairwell, terminating in a 12- by 12- by 6-in. deep junction box in ceiling structure of second floor. Run ½-in. EMT from junction box to a point above each wall heater and terminate in a 4-in.-square, 1½-in.-deep junction box with blank cover. Run No. 700 Wiremold (on wall surface) from junction box to wall heater.

See detail in Fig. 8-8.

The panelboard schedule appears in Fig. 8-9.

Many electrical designers and contractors are required to calculate the heat loss in residences because more and more homes now utilize baseboard electric heaters as their only source of heat. Since there have been many methods previously described for calculating heat loss, it will not be covered in this chapter. However, a good reference for residential heat-loss calculations is the book *Making Electrical Calculations* by J. F.

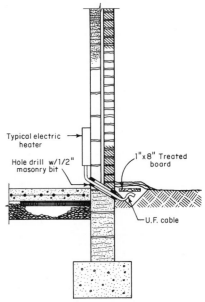

Fig. 8-7 Electrical detail indicating method of installing cable in a shallow trench outside of building.

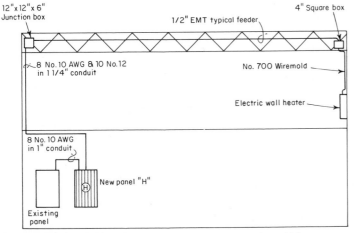

Fig. 8-8 Electrical detail showing method of connecting Wiremold from a junction box to a wall heater.

Note: Provide and install five 30-A 2-pole circuit breakers.

PANELBOARD SCHEDULE										
Panel No.	Type Cabinet	Panel Mains			Branches					Item Fed
		Amperes	Volts	Phase	1P	2P	3P	Prot	Frame	
H	Surface	200	120/240	3∅		21		20A	100A	Heaters
	Square "D" Type NQO									

Fig. 8-9 Panelboard schedule.

McPartland, published in 1975 by *Electrical Construction and Maintenance* Magazine, New York.

Three working drawings using electric heat are shown in Figs. 8-10, 8-11, and 8-12. Related wiring, location of the heaters,

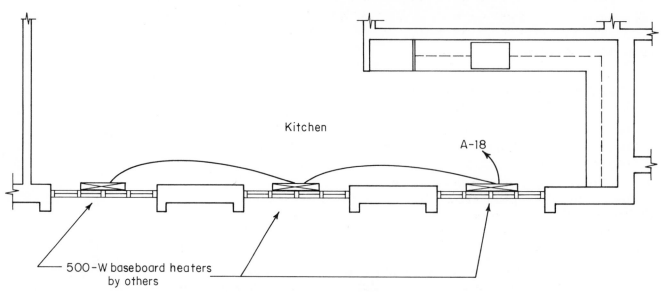

Fig. 8-10 How electric-heating equipment is shown on a working drawing.

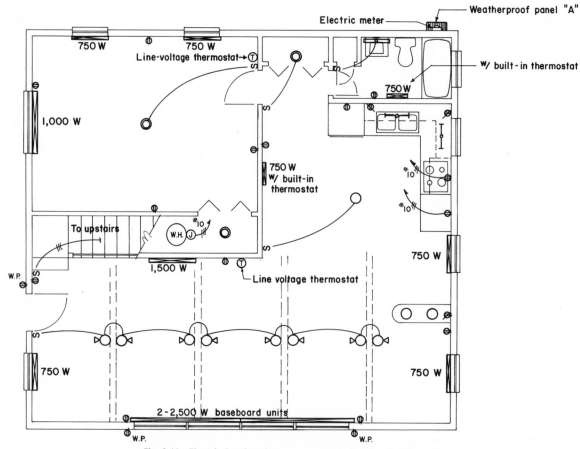

Fig. 8-11 Electric heating shown on a working drawing, first-floor plan.

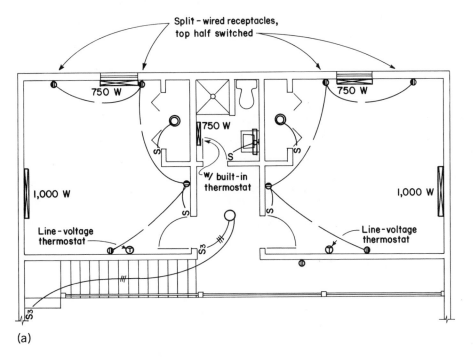

(a)

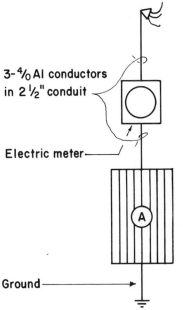

3-⁴⁄₀ Al conductors
in 2 ½" conduit

Electric meter

Ground

Weatherproof 200-A I∮
panel to contain:

1-15 A ground fault protector for outside receptacles.
2-20A IP CB for appliance circuits.
3-15A IP CB for receptacles.
3-15A IP CB for lights.
1-30A 2P CB for cooking top.
1-30A 2P CB for oven.
1-30A 2P CB for water heater.
10-20A 2P CB for electric heat.
5-Spares

(b)

Fig. 8-12 (a) Electric heating shown on a working drawing, second-floor plan corresponding to plan in Fig. 8-11; (b) power-riser diagram for the entire building.

etc., are shown on the floor plan. The actual working drawings also used panelboard schedules and electric heat schedules to give further details of construction.

Chapter 9

Service Equipment

In the order of passage of electrical energy from the power company's lines to the point of first overcurrent protection and disconnecting means in the building, these are the following generally recognized sections of an electrical system:

Service Drop: The overhead conductors, through which electrical service is supplied, between the last power company pole and the point of their connection to the service facilities located at the building or other support used for the purpose

Service Entrance: All components between the point of termination of the overhead service drop or underground service lateral and the building main disconnecting device, with the exception of the power company's metering equipment

Service-Entrance Conductors: The conductors between the point of termination of the overhead service drop or underground service lateral and the main disconnecting device in the building

Service Lateral: The underground conductors, through which service is supplied, between the power company's distribution facilities and the first point of their connection to the building or area service facilities located at the building or other support used for the purpose

Service-Entrance Equipment: Provides overcurrent protection to the feeder and service conductors; provides a means of disconnecting the feeders from energized service conductors; and provides a means of measuring the energy used by the use of metering equipment

AVAILABLE VOLTAGES

The power company usually supplies alternating current at a frequency of 60 hertz (Hz), or cycles per second. The standard voltages which are normally available are listed in the following table.

Voltage	Phase	Wires
120/240	1	3
120/208	3	4 wye
120/240	3	4 delta
277/480	3	4 wye
2,400	3	3 delta
4,160	3	4 wye
12,470	3	4 wye
34,500	3	3

The availability of the above voltages is dependent upon the power company's available facilities, method of service, and the size and character of the customer's load.

Service for small lighting, heating, and power installations will normally be supplied as single-phase, 120/240-V, three-wire.

When three-phase service is required, 120/208-V or 277/480-V wye is usually preferred over delta as the "red-leg" phase of the delta prevents proper balancing of 120-V loads.

Three-phase service may also be made available at 2,400, 4,160, 12,470 and 34,500 V. Service at these voltages is for large power installations and is normally provided from an individual customer substation. When service is taken at 2,400 or 4,160 V, it is usually necessary for the consulting engineer to design the 15-kV class equipment, which includes underground cable, cutouts, arresters, primary breakers, primary terminations, transformer vaults, etc.

As mentioned previously, most services are delivered at 120/240-V single-phase or 102/208-V three-phase. However, 277/480-V services are quite common in the larger buildings today. In such cases, most of the power equipment, such as heating and cooling equipment, pumps, motors, etc., are rated for 480-V operation. Convenience receptacles, however, along with some possible lighting circuits, will require 120 V. In such instances, stepdown transformers are used to transform the voltage from 277/480 V of the primary circuit to 120/208 V for the secondary section. These transformers are usually of the "dry" type, and the capacity or size will depend upon the load requirements.

Overhead Service Installations

A safe, substantial support is necessary for the power company's overhead service. When designing or specifying any electrical service, requirements set forth in the National Electrical Code must be followed as well as the requirements of the local power company.

Figures 9-1 through 9-11 show various overhead services and meter installations that are acceptable to most power companies. Other examples of services which are of a higher current-carrying capacity, as well as practical applications of all types, are given later in this chapter.

**TYPICAL 200 AMP. OR LESS
OVERHEAD SERVICE INSTALLATION**

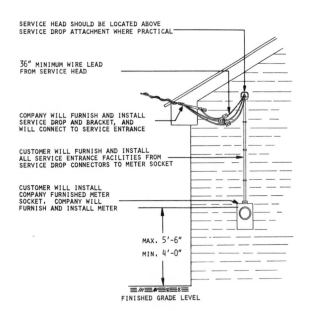

NOTES:

1. CUSTOMER MUST CONSULT WITH COMPANY FOR POINT OF ATTACHMENT OF SERVICE DROP AND METER LOCATION.

2. ALL CUSTOMER WORK MUST BE COMPLETED AND REQUIRED INSPECTIONS BE OBTAINED BEFORE COMPANY WILL PROVIDE SERVICE.

3. SERVICE DROP POINT OF ATTACHMENT MUST BE OF SUFFICIENT HEIGHT TO ALLOW FOLLOWING MINIMUM SERVICE DROP CLEARANCES:

   (A) TEN FEET (10') FOR MULTIPLEX SERVICES ABOVE SIDEWALKS AND FINISHED GRADES.

   (B) TWELVE FEET (12') FOR OPEN WIRE SERVICES ABOVE SIDEWALKS AND FINISHED GRADES.

   (C) TWELVE FEET (12') OVER RESIDENTIAL DRIVEWAYS.

   (D) EIGHTEEN FEET (18') OVER PUBLIC STREETS, ALLEYS, PUBLIC AND SHOPPING CENTER PARKING LOTS.

   (E) TWO FEET (2') CLEARANCE FROM TELEPHONE AND CATV WIRES AT MIDSPAN CROSSINGS.

4. INHIBITOR COMPOUND SHALL BE USED ON ALL ALUMINUM WIRE TERMINATIONS.

(8/73)

Fig. 9-1

**TYPICAL 200 AMP. OR LESS
OVERHEAD SERVICE MAST INSTALLATION**

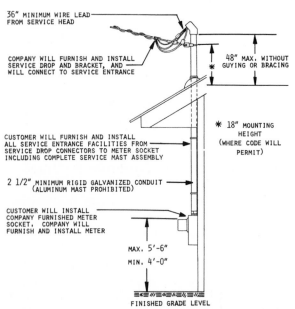

NOTES:

1. CUSTOMER MUST CONSULT WITH COMPANY FOR POINT OF ATTACHMENT OF SERVICE DROP AND METER LOCATION.

2. ALL CUSTOMER WORK MUST BE COMPLETED AND REQUIRED INSPECTIONS BE OBTAINED BEFORE COMPANY WILL PROVIDE SERVICE.

3. SERVICE DROP POINT OF ATTACHMENT MUST BE OF SUFFICIENT HEIGHT TO ALLOW FOLLOWING MINIMUM SERVICE DROP CLEARANCES:

   (A) TEN FEET (10') FOR MULTIPLEX SERVICES ABOVE SIDEWALKS AND FINISHED GRADES.

   (B) TWELVE FEET (12') FOR OPEN WIRE SERVICES ABOVE SIDEWALKS AND FINISHED GRADES.

   (C) TWELVE FEET (12') OVER RESIDENTIAL DRIVEWAYS.

   (D) EIGHTEEN FEET (18') OVER PUBLIC STREETS, ALLEYS, PUBLIC AND SHOPPING CENTER PARKING LOTS.

   (E) TWO FEET (2') CLEARANCE FROM TELEPHONE AND CATV WIRES AT MIDSPAN CROSSINGS.

4. INHIBITOR COMPOUND SHALL BE USED ON ALL ALUMINUM WIRE TERMINATIONS.

(8/73)

Fig. 9-2

TYPICAL 200 AMP. OR LESS
PERMANENT OVERHEAD SERVICE POLE
(PERMANENT FEEDER)

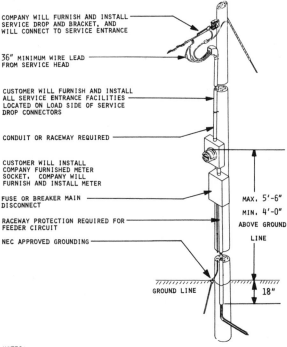

COMPANY WILL FURNISH AND INSTALL
SERVICE DROP AND BRACKET, AND
WILL CONNECT TO SERVICE ENTRANCE

36" MINIMUM WIRE LEAD
FROM SERVICE HEAD

CUSTOMER WILL FURNISH AND INSTALL
ALL SERVICE ENTRANCE FACILITIES
LOCATED ON LOAD SIDE OF SERVICE
DROP CONNECTORS

CONDUIT OR RACEWAY REQUIRED

CUSTOMER WILL INSTALL
COMPANY FURNISHED METER
SOCKET. COMPANY WILL
FURNISH AND INSTALL METER

FUSE OR BREAKER MAIN
DISCONNECT

RACEWAY PROTECTION REQUIRED FOR
FEEDER CIRCUIT

NEC APPROVED GROUNDING

MAX. 5'-6"

MIN. 4'-0"

ABOVE GROUND

LINE

GROUND LINE

18"

NOTES:

1. CUSTOMER MUST CONSULT WITH COMPANY FOR LOCATION OF SUCH POLE
   REQUIRED WHEN INSTALLATION OF METER ON BUILDING NOT FEASIBLE
   (MOBILE HOME, ETC.).

2. ALL CUSTOMER WORK MUST BE COMPLETED AND REQUIRED INSPECTIONS
   BE OBTAINED BEFORE COMPANY WILL PROVIDE SERVICE.

3. CUSTOMER WILL FURNISH AND INSTALL COMPANY APPROVED POLE, TREATED
   WITH CHEMICAL PRESERVATIVE. MINIMUM REQUIREMENTS FOR POLE: 25'
   LENGTH CLASS 6 (17" MIN. TOP CIRCUM., 23" MIN. CIRCUM. AT 6'
   FROM BUTT); INSTALLED MIN. 4 1/2' IN GROUND WELL TAMPED AND
   GUYED AGAINST PULL OF SERVICE DROP IF DROP LENGTH EXCEEDS 50'.

4. ANTENNAS OR AERIALS SHALL NOT BE ATTACHED TO ANY POLE
   USED FOR SUPPLYING ELECTRIC SERVICE TO THE CUSTOMER.

5. INHIBITOR COMPOUND SHALL BE USED ON ALL ALUMINUM WIRE TERMINATIONS.

(8/73)

Fig. 9-3

TYPICAL 200 AMP. OR LESS
PERMANENT OVERHEAD SERVICE POLE
(WEATHERPROOF POWER OUTLET)

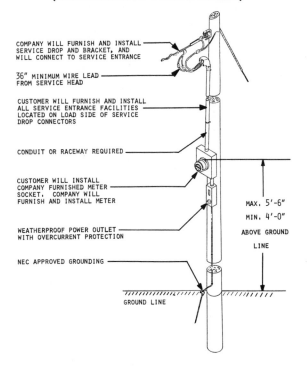

COMPANY WILL FURNISH AND INSTALL
SERVICE DROP AND BRACKET, AND
WILL CONNECT TO SERVICE ENTRANCE

36" MINIMUM WIRE LEAD
FROM SERVICE HEAD

CUSTOMER WILL FURNISH AND INSTALL
ALL SERVICE ENTRANCE FACILITIES
LOCATED ON LOAD SIDE OF SERVICE
DROP CONNECTORS

CONDUIT OR RACEWAY REQUIRED

CUSTOMER WILL INSTALL
COMPANY FURNISHED METER
SOCKET. COMPANY WILL
FURNISH AND INSTALL METER

WEATHERPROOF POWER OUTLET
WITH OVERCURRENT PROTECTION

NEC APPROVED GROUNDING

MAX. 5'-6"

MIN. 4'-0"

ABOVE GROUND

LINE

GROUND LINE

NOTES:

1. CUSTOMER MUST CONSULT WITH COMPANY FOR LOCATION OF SUCH POLE
   REQUIRED WHEN INSTALLATION OF METER ON BUILDING NOT FEASIBLE
   (MOBILE HOME, ETC.).

2. ALL CUSTOMER WORK MUST BE COMPLETED AND REQUIRED INSPECTIONS
   BE OBTAINED BEFORE COMPANY WILL PROVIDE SERVICE.

3. CUSTOMER WILL FURNISH AND INSTALL COMPANY APPROVED POLE, TREATED
   WITH CHEMICAL PRESERVATIVE. MINIMUM REQUIREMENTS FOR POLE: 25'
   LENGTH, CLASS 6 (17" MIN. TOP CIRCUM., 23" MIN. CIRCUM. AT 6'
   FROM BUTT); INSTALLED MIN. 4 1/2' IN GROUND, WELL TAMPED AND
   GUYED AGAINST PULL OF SERVICE DROP IF DROP LENGTH EXCEEDS 50'.

4. ANTENNAS OR AERIALS SHALL NOT BE ATTACHED TO ANY POLE
   USED FOR SUPPLYING ELECTRIC SERVICE TO THE CUSTOMER.

5. INHIBITOR COMPOUND SHALL BE USED ON ALL ALUMINUM WIRE TERMINATIONS.

(8/73)

Fig. 9-4

CENTRAL FARM POLE SERVICE

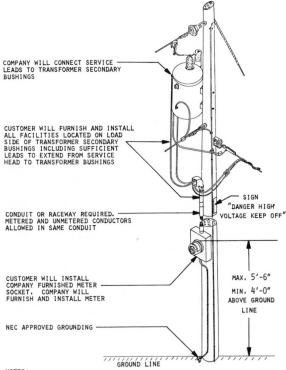

COMPANY WILL CONNECT SERVICE
LEADS TO TRANSFORMER SECONDARY
BUSHINGS

CUSTOMER WILL FURNISH AND INSTALL
ALL FACILITIES LOCATED ON LOAD
SIDE OF TRANSFORMER SECONDARY
BUSHINGS INCLUDING SUFFICIENT
LEADS TO EXTEND FROM SERVICE
HEAD TO TRANSFORMER BUSHINGS

SIGN
"DANGER HIGH
VOLTAGE KEEP OFF"

CONDUIT OR RACEWAY REQUIRED.
METERED AND UNMETERED CONDUCTORS
ALLOWED IN SAME CONDUIT

CUSTOMER WILL INSTALL
COMPANY FURNISHED METER
SOCKET. COMPANY WILL
FURNISH AND INSTALL METER

MAX. 5'-6"
MIN. 4'-0"
ABOVE GROUND
LINE

NEC APPROVED GROUNDING

GROUND LINE

NOTES:

1. CUSTOMER MUST CONSULT WITH COMPANY BEFORE INSTALLING CENTRAL
   FARM POLE SERVICE.

2. CUSTOMER MUST NOTIFY COMPANY AND OBTAIN ITS PERMISSION BEFORE
   PERFORMING ANY WORK ON FACILITIES LOCATED ON THIS POLE.

3. COMPANY WILL FURNISH AND INSTALL POLE WHEN NECESSARY TO SUPPORT
   TRANSFORMER.

4. NOT MORE THAN FIVE (5) CUSTOMER FEEDER CIRCUITS WILL BE PERMITTED
   ON COMPANY OWNED POLE. ALL FEEDER CIRCUITS MUST BE MULTIPLEX.

5. INHIBITOR COMPOUND SHALL BE USED ON ALL ALUMINUM WIRE TERMINATIONS.

6. WHEN LOAD EXCEEDS 200 AMP., METERING TRANSFORMERS WILL BE
   LOCATED AT TOP OF POLE WITH ALL METERING FACILITIES FURNISHED
   AND INSTALLED BY COMPANY.

(8/73)

Fig. 9-5

200 AMP. OR LESS
CENTRAL FARM POLE SERVICE
(EMERGENCY STAND-BY GENERATOR FACILITIES)

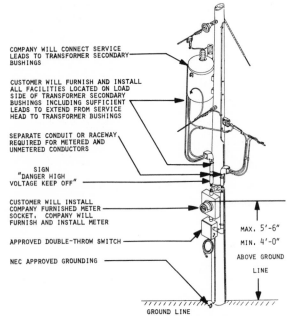

COMPANY WILL CONNECT SERVICE
LEADS TO TRANSFORMER SECONDARY
BUSHINGS

CUSTOMER WILL FURNISH AND INSTALL
ALL FACILITIES LOCATED ON LOAD
SIDE OF TRANSFORMER SECONDARY
BUSHINGS INCLUDING SUFFICIENT
LEADS TO EXTEND FROM SERVICE
HEAD TO TRANSFORMER BUSHINGS

SEPARATE CONDUIT OR RACEWAY
REQUIRED FOR METERED AND
UNMETERED CONDUCTORS

SIGN
"DANGER HIGH
VOLTAGE KEEP OFF"

CUSTOMER WILL INSTALL
COMPANY FURNISHED METER
SOCKET. COMPANY WILL
FURNISH AND INSTALL METER

MAX. 5'-6"
MIN. 4'-0"
ABOVE GROUND
LINE

APPROVED DOUBLE-THROW SWITCH

NEC APPROVED GROUNDING

GROUND LINE

NOTES:

1. CUSTOMER MUST CONSULT WITH COMPANY BEFORE INSTALLING EMERGENCY
   STAND-BY GENERATOR FACILITIES AND MUST ALSO ENTER INTO A
   COMPANY INDEMNIFICATION AND SAVE-HARMLESS AGREEMENT.

2. ALL EMERGENCY STAND-BY GENERATOR FACILITIES MUST BE INSTALLED
   IN ACCORDANCE WITH APPLICABLE CODES AND BE SUBJECT TO COMPANY
   APPROVAL.

3. CUSTOMER MUST NOTIFY COMPANY AND OBTAIN ITS PERMISSION BEFORE
   PERFORMING ANY WORK ON FACILITIES LOCATED ON THIS POLE.

4. COMPANY WILL FURNISH AND INSTALL POLE WHEN NECESSARY TO
   SUPPORT TRANSFORMER.

5. NOT MORE THAN FIVE (5) CUSTOMER FEEDER CIRCUITS WILL BE PERMITTED
   ON COMPANY OWNED POLE. ALL FEEDER CIRCUITS MUST BE MULTIPLEX.

6. ALL CONDUITS MUST BE INSTALLED ON SAME SIDE OF POLE IN ORDER
   TO PROVIDE CLIMBING SPACE.

7. INHIBITOR COMPOUND SHALL BE USED ON ALL ALUMINUM WIRE TERMINATIONS.

(8/73)

Fig. 9-6

400 AMP. CENTRAL FARM POLE SERVICE
(EMERGENCY STAND-BY GENERATOR FACILITIES
WITH POLE TOP SWITCH)

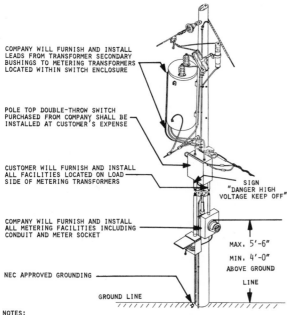

COMPANY WILL FURNISH AND INSTALL LEADS FROM TRANSFORMER SECONDARY BUSHINGS TO METERING TRANSFORMERS LOCATED WITHIN SWITCH ENCLOSURE

POLE TOP DOUBLE-THROW SWITCH PURCHASED FROM COMPANY SHALL BE INSTALLED AT CUSTOMER'S EXPENSE

CUSTOMER WILL FURNISH AND INSTALL ALL FACILITIES LOCATED ON LOAD SIDE OF METERING TRANSFORMERS

SIGN "DANGER HIGH VOLTAGE KEEP OFF"

COMPANY WILL FURNISH AND INSTALL ALL METERING FACILITIES INCLUDING CONDUIT AND METER SOCKET

MAX. 5'-6"

MIN. 4'-0"

ABOVE GROUND LINE

NEC APPROVED GROUNDING

GROUND LINE

NOTES:

1. CUSTOMER MUST CONSULT WITH COMPANY BEFORE INSTALLING EMERGENCY STAND-BY GENERATOR FACILITIES AND MUST ALSO ENTER INTO A COMPANY INDEMNIFICATION AND SAVE-HARMLESS AGREEMENT.

2. ALL EMERGENCY STAND-BY GENERATOR FACILITIES MUST BE INSTALLED IN ACCORDANCE WITH APPLICABLE CODES AND BE SUBJECT TO COMPANY APPROVAL.

3. CUSTOMER MUST NOTIFY COMPANY AND OBTAIN ITS PERMISSION BEFORE PERFORMING ANY WORK ON FACILITIES LOCATED ON THIS POLE.

4. DRAWING SHOWS USE OF POLE TOP SWITCH. CUSTOMER TO CONSULT COMPANY FOR ALTERNATE LOCATIONS.

5. MAINTENANCE OF POLE TOP SWITCH WILL BE PERFORMED BY COMPANY AT CUSTOMER'S EXPENSE.

6. COMPANY WILL FURNISH AND INSTALL POLE WHEN NECESSARY TO SUPPORT TRANSFORMER.

7. ALL CONDUITS AND CONTROL ROD MUST BE INSTALLED ON SAME SIDE OF POLE IN ORDER TO PROVIDE CLIMBING SPACE.

8. INHIBITOR COMPOUND SHALL BE USED ON ALL ALUMINUM WIRE TERMINATIONS.

(8/73)

Fig. 9-7

TEMPORARY OVERHEAD
SERVICE SUPPORT

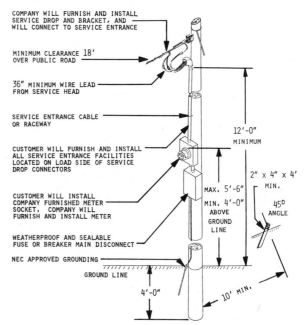

COMPANY WILL FURNISH AND INSTALL SERVICE DROP AND BRACKET, AND WILL CONNECT TO SERVICE ENTRANCE

MINIMUM CLEARANCE 18' OVER PUBLIC ROAD

36" MINIMUM WIRE LEAD FROM SERVICE HEAD

SERVICE ENTRANCE CABLE OR RACEWAY

12'-0" MINIMUM

CUSTOMER WILL FURNISH AND INSTALL ALL SERVICE ENTRANCE FACILITIES LOCATED ON LOAD SIDE OF SERVICE DROP CONNECTORS

2" x 4" x 4' MIN.

45° ANGLE

CUSTOMER WILL INSTALL COMPANY FURNISHED METER SOCKET. COMPANY WILL FURNISH AND INSTALL METER

MAX. 5'-6"

MIN. 4'-0" ABOVE GROUND LINE

WEATHERPROOF AND SEALABLE FUSE OR BREAKER MAIN DISCONNECT

NEC APPROVED GROUNDING

GROUND LINE

4'-0"

10' MIN.

NOTES:

1. CUSTOMER MUST CONSULT WITH COMPANY FOR LOCATION OF TEMPORARY SERVICE POLE OR SUPPORT; ALSO, CUSTOMER MUST PAY APPLICABLE CHARGES.

2. ALL CUSTOMER WORK MUST BE COMPLETED AND REQUIRED INSPECTIONS BE OBTAINED BEFORE COMPANY WILL PROVIDE SERVICE.

3. SERVICE POLE OR SUPPORT, FURNISHED AND INSTALLED BY CUSTOMER. REQUIREMENTS: 18' MINIMUM LENGTH CHEMICAL PRESERVATIVE TREATED POLE (MIN. CLASS 9, OR 5" DIAMETER AT TOP); OR 18' MINIMUM CONTINUOUS LENGTH OF 4" x 4" REDWOOD, OR OTHER CHEMICAL PRESERVATIVE TREATED POST FREE OF UNACCEPTABLE DEFECTS.

4. GUY WIRE FURNISHED AND INSTALLED BY CUSTOMER, IF COMPANY SERVICE DROP LENGTH EXCEEDS 50'. REQUIREMENTS: #9 IRON FENCE WIRE, OR #7 ALUMINUM CLOTHES LINE WIRE, OR TV TOWER GUY WIRE, INSTALLED AGAINST PULL OF SERVICE DROP. WOOD BRACING FOR POLE OR SUPPORT NOT ACCEPTABLE.

5. INHIBITOR COMPOUND SHALL BE USED ON ALL ALUMINUM WIRE TERMINATIONS.

6. COMPLETED CUSTOMER INSTALLATION SUBJECT TO COMPANY APPROVAL PRIOR TO CONNECTION.

(8/73)

Fig. 9-8

MOBILE HOME PARK SERVICE POLE/METER INSTALLATION

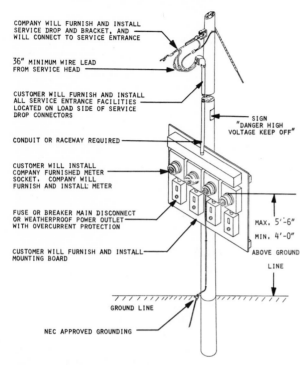

COMPANY WILL FURNISH AND INSTALL SERVICE DROP AND BRACKET, AND WILL CONNECT TO SERVICE ENTRANCE

36" MINIMUM WIRE LEAD FROM SERVICE HEAD

CUSTOMER WILL FURNISH AND INSTALL ALL SERVICE ENTRANCE FACILITIES LOCATED ON LOAD SIDE OF SERVICE DROP CONNECTORS

SIGN "DANGER HIGH VOLTAGE KEEP OFF"

CONDUIT OR RACEWAY REQUIRED

CUSTOMER WILL INSTALL COMPANY FURNISHED METER SOCKET. COMPANY WILL FURNISH AND INSTALL METER

FUSE OR BREAKER MAIN DISCONNECT OR WEATHERPROOF POWER OUTLET WITH OVERCURRENT PROTECTION

CUSTOMER WILL FURNISH AND INSTALL MOUNTING BOARD

MAX. 5'-6"

MIN. 4'-0"

ABOVE GROUND LINE

GROUND LINE

NEC APPROVED GROUNDING

NOTES:

1. CUSTOMER MUST CONSULT WITH COMPANY FOR METHOD OF SERVING MOBILE HOME PARK AND LOCATION OF SERVICE POLES.

2. CUSTOMER MUST NOTIFY COMPANY AND OBTAIN ITS PERMISSION BEFORE PERFORMING ANY WORK ON FACILITIES ON THIS POLE.

3. ALL CUSTOMER WORK MUST BE COMPLETED AND REQUIRED INSPECTIONS BE OBTAINED BEFORE COMPANY WILL PROVIDE SERVICE.

4. COMPANY WILL FURNISH AND INSTALL POLE WHEN A MINIMUM OF TWO (2) AND A MAXIMUM OF FOUR (4) METERS ARE INSTALLED ON THE POLE. A MOBILE HOME PARK SHOULD PREFERABLY BE DESIGNED TO SPECIFY FOUR METERS PER POLE.

5. FOUR-METER MOUNTING BOARD REQUIREMENTS: 60" WIDE, 48" HIGH, 3/4" THICK WITH 2" X 4" BATTENS; PAINTED AND SECURELY ATTACHED TO POLE WITH 1/2" THRU BOLTS.

6. INHIBITOR COMPOUND SHALL BE USED ON ALL ALUMINUM WIRE TERMINATIONS.

(8/73)

Fig. 9-9

TYPICAL OVERHEAD TRANSFORMER – OPERATED METER INSTALLATION

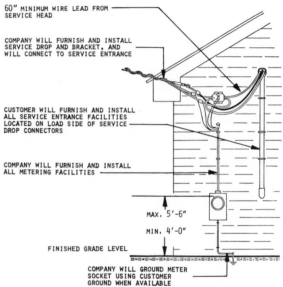

60" MINIMUM WIRE LEAD FROM SERVICE HEAD

COMPANY WILL FURNISH AND INSTALL SERVICE DROP AND BRACKET, AND WILL CONNECT TO SERVICE ENTRANCE

CUSTOMER WILL FURNISH AND INSTALL ALL SERVICE ENTRANCE FACILITIES LOCATED ON LOAD SIDE OF SERVICE DROP CONNECTORS

COMPANY WILL FURNISH AND INSTALL ALL METERING FACILITIES

MAX. 5'-6"

MIN. 4'-0"

FINISHED GRADE LEVEL

COMPANY WILL GROUND METER SOCKET USING CUSTOMER GROUND WHEN AVAILABLE

NOTES:

1. CUSTOMER MUST CONSULT WITH COMPANY FOR POINT OF ATTACHMENT OF SERVICE DROP AND METER LOCATION.

2. CUSTOMER TO CONSULT COMPANY FOR THREE-PHASE, OR POLE MOUNTED, TRANSFORMER OPERATED METER INSTALLATION.

3. ALL CUSTOMER WORK MUST BE COMPLETED AND REQUIRED INSPECTIONS BE OBTAINED BEFORE COMPANY WILL PROVIDE SERVICE.

4. SERVICE DROP POINT OF ATTACHMENT MUST BE OF SUFFICIENT HEIGHT TO ALLOW FOLLOWING MINIMUM SERVICE DROP CLEARANCES:

   (A) TEN FEET (10') FOR MULTIPLEX SERVICES ABOVE SIDEWALKS AND FINISHED GRADES.

   (B) TWELVE FEET (12') FOR OPEN WIRE SERVICES ABOVE SIDEWALKS AND FINISHED GRADES.

   (C) TWELVE FEET (12') OVER RESIDENTIAL DRIVEWAYS.

   (D) EIGHTEEN FEET (18') OVER PUBLIC STREETS, ALLEYS, PUBLIC AND SHOPPING CENTER PARKING LOTS.

   (E) TWO FEET (2') CLEARANCE FROM TELEPHONE AND CATV WIRES AT MIDSPAN CROSSINGS.

5. INHIBITOR COMPOUND SHALL BE USED ON ALL ALUMINUM WIRE TERMINATIONS.

(8/73)

Fig. 9-10

**TYPICAL OVERHEAD MULTIPLE METER
INSTALLATION**

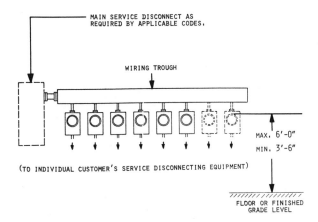

NOTES:

1. CUSTOMER MUST CONSULT WITH COMPANY FOR POINT OF ATTACHMENT OF
   SERVICE DROP AND METERING LOCATION.

2. ALL CUSTOMER WORK MUST BE COMPLETED AND REQUIRED INSPECTIONS
   BE OBTAINED BEFORE COMPANY WILL PROVIDE SERVICE.

3. THE CUSTOMER SHALL BE RESPONSIBLE FOR FURNISHING AND INSTALLING
   ALL SERVICE ENTRANCE WIRING AND FACILITIES FROM THE COMPANY'S
   SERVICE DROP TO THE METER SOCKETS.

4. CUSTOMER WILL INSTALL COMPANY FURNISHED METER SOCKETS.

5. COMPANY WILL FURNISH AND INSTALL METERS.

6. METERED CONDUCTORS SHALL NOT BE INSTALLED IN WIRING TROUGH(S).

7. CUSTOMER MAY INSTALL METER STACK OR METER TROUGH TYPE
   EQUIPMENT, SUBJECT TO COMPANY APPROVAL, IN LIEU OF COMPANY
   FURNISHED METER SOCKETS.

8. WHEN THE SERVICE ENTRANCE CONSISTS OF MORE THAN ONE SET OF
   CONDUCTORS, THE INDIVIDUAL LOADS MUST BE CONNECTED SO AS TO
   BE BALANCED AMONG ALL SETS OF CONDUCTORS.

9. WIRING TROUGH(S) AND MAIN SERVICE DISCONNECT
   SHALL BE SEALABLE AND SHALL ALSO BE WEATHERPROOF WHEN
   INSTALLED OUTDOORS.

10. INHIBITOR COMPOUND SHALL BE USED ON ALL ALUMINUM WIRE TERMINATIONS.

(8/73)

Fig. 9-11

Underground Services

Figures 9-12 through 9-15 show various underground service
and meter installations accepted by most power companies.
Other examples are also shown later in this chapter.

Service-entrance equipment will usually consist of either a
factory-built switchboard with all switches, fuses, or circuit
breakers, and, in some instances, the metering equipment built
in, or an assembly of individual units of switches or circuit
breakers, metering equipment, and auxiliary gutters.

In the design of service-entrance equipment, the following
factors must be taken into consideration:

1. Service characteristics available from the power company
2. Total connected electrical load
3. Total demand of electrical load

4. Capacity for present and future requirements
5. Requirements of National Electrical Code, local ordinances, and local power company
6. Type and cost of equipment
7. Delivery date of equipment
8. Physical dimensions of equipment and space available in building

TYPICAL 200 AMP. OR LESS
UNDERGROUND SERVICE INSTALLATION

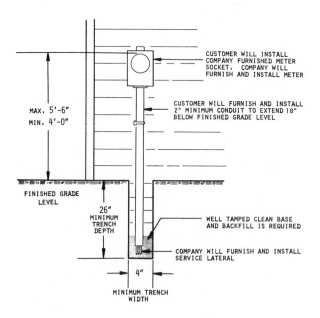

CUSTOMER WILL INSTALL COMPANY FURNISHED METER SOCKET. COMPANY WILL FURNISH AND INSTALL METER

CUSTOMER WILL FURNISH AND INSTALL 2" MINIMUM CONDUIT TO EXTEND 18" BELOW FINISHED GRADE LEVEL

MAX. 5'-6"
MIN. 4'-0"

FINISHED GRADE LEVEL

26" MINIMUM TRENCH DEPTH

WELL TAMPED CLEAN BASE AND BACKFILL IS REQUIRED

COMPANY WILL FURNISH AND INSTALL SERVICE LATERAL

4"

MINIMUM TRENCH WIDTH

NOTES:

1. CUSTOMER MUST CONSULT WITH COMPANY FOR POINT OF ATTACHMENT OF SERVICE LATERAL AND METER LOCATION; ALSO CUSTOMER MUST PAY ANY APPLICABLE CHARGES OR DEPOSITS ASSOCIATED WITH UNDERGROUND SERVICE BEFORE CONSTRUCTION IS BEGUN.

2. ALL CUSTOMER WORK MUST BE COMPLETED AND REQUIRED INSPECTIONS BE OBTAINED BEFORE COMPANY WILL PROVIDE SERVICE.

3. INHIBITOR COMPOUND SHALL BE USED ON ALL ALUMINUM WIRE TERMINATIONS.

(8/73)

Fig. 9-12

TYPICAL 200 AMP. OR LESS
PERMANENT UNDERGROUND SERVICE SUPPORT
(PERMANENT FEEDER)

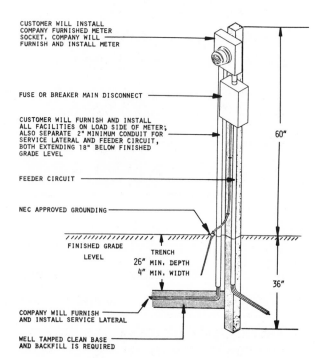

CUSTOMER WILL INSTALL COMPANY FURNISHED METER SOCKET. COMPANY WILL FURNISH AND INSTALL METER

FUSE OR BREAKER MAIN DISCONNECT

CUSTOMER WILL FURNISH AND INSTALL ALL FACILITIES ON LOAD SIDE OF METER; ALSO SEPARATE 2" MINIMUM CONDUIT FOR SERVICE LATERAL AND FEEDER CIRCUIT, BOTH EXTENDING 18" BELOW FINISHED GRADE LEVEL

FEEDER CIRCUIT

NEC APPROVED GROUNDING

FINISHED GRADE LEVEL

TRENCH
26" MIN. DEPTH
4" MIN. WIDTH

60"

36"

COMPANY WILL FURNISH AND INSTALL SERVICE LATERAL

WELL TAMPED CLEAN BASE AND BACKFILL IS REQUIRED

NOTES:

1. CUSTOMER MUST CONSULT WITH COMPANY FOR LOCATION OF SUCH SUPPORT REQUIRED WHEN INSTALLATION OF METER ON BUILDING NOT FEASIBLE (MOBILE HOME, ETC.); ALSO CUSTOMER MUST PAY ANY APPLICABLE CHARGES OR DEPOSITS ASSOCIATED WITH UNDERGROUND SERVICE BEFORE CONSTRUCTION IS BEGUN.

2. ALL CUSTOMER WORK MUST BE COMPLETED AND REQUIRED INSPECTIONS BE OBTAINED BEFORE COMPANY WILL PROVIDE SERVICE.

3. CUSTOMER WILL FURNISH AND INSTALL A NOMINAL 4" x 4" x 8' REDWOOD OR OTHER CHEMICAL PRESERVATIVE TREATED POST OR OTHER COMPANY APPROVED SUPPORT.

4. INHIBITOR COMPOUND SHALL BE USED ON ALL ALUMINUM WIRE TERMINATIONS.

(8/73)

Fig. 9-13

TYPICAL 200 AMP. OR LESS
PERMANENT UNDERGROUND SERVICE SUPPORT
(WEATHERPROOF POWER OUTLET)

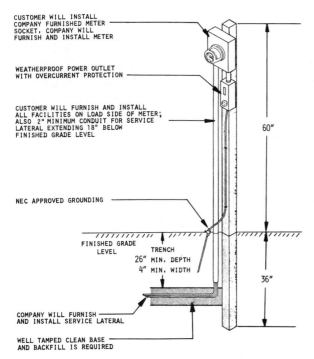

CUSTOMER WILL INSTALL
COMPANY FURNISHED METER
SOCKET, COMPANY WILL
FURNISH AND INSTALL METER

WEATHERPROOF POWER OUTLET
WITH OVERCURRENT PROTECTION

CUSTOMER WILL FURNISH AND INSTALL
ALL FACILITIES ON LOAD SIDE OF METER;
ALSO 2" MINIMUM CONDUIT FOR SERVICE
LATERAL EXTENDING 18" BELOW
FINISHED GRADE LEVEL

60"

NEC APPROVED GROUNDING

FINISHED GRADE
LEVEL

TRENCH
26" MIN. DEPTH
4" MIN. WIDTH

36"

COMPANY WILL FURNISH
AND INSTALL SERVICE LATERAL

WELL TAMPED CLEAN BASE
AND BACKFILL IS REQUIRED

NOTES:

1. CUSTOMER MUST CONSULT WITH COMPANY FOR LOCATION OF SUCH SUPPORT
REQUIRED WHEN INSTALLATION OF METER ON BUILDING NOT FEASIBLE
(MOBILE HOME, ETC.); ALSO CUSTOMER MUST PAY ANY APPLICABLE
CHARGES OR DEPOSITS ASSOCIATED WITH UNDERGROUND SERVICE BEFORE
CONSTRUCTION IS BEGUN.

2. ALL CUSTOMER WORK MUST BE COMPLETED AND REQUIRED INSPECTIONS
BE OBTAINED BEFORE COMPANY WILL PROVIDE SERVICE.

3. CUSTOMER WILL FURNISH AND INSTALL A NOMINAL 4" x 4" x 8' REDWOOD
OR OTHER CHEMICAL PRESERVATIVE TREATED POST OR OTHER COMPANY
APPROVED SUPPORT.

4. INHIBITOR COMPOUND SHALL BE USED ON ALL ALUMINUM WIRE TERMINATIONS.

(8/73)

Fig. 9-14

TEMPORARY UNDERGROUND
SERVICE SUPPORT

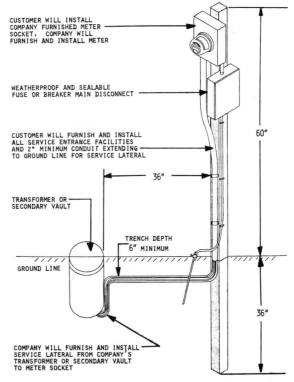

CUSTOMER WILL INSTALL
COMPANY FURNISHED METER
SOCKET. COMPANY WILL
FURNISH AND INSTALL METER

WEATHERPROOF AND SEALABLE
FUSE OR BREAKER MAIN DISCONNECT

CUSTOMER WILL FURNISH AND INSTALL
ALL SERVICE ENTRANCE FACILITIES
AND 2" MINIMUM CONDUIT EXTENDING
TO GROUND LINE FOR SERVICE LATERAL

60"

36"

TRANSFORMER OR
SECONDARY VAULT

TRENCH DEPTH
6" MINIMUM

GROUND LINE

36"

COMPANY WILL FURNISH AND INSTALL
SERVICE LATERAL FROM COMPANY'S
TRANSFORMER OR SECONDARY VAULT
TO METER SOCKET

NOTES:

1. CUSTOMER MUST CONSULT WITH COMPANY FOR LOCATION OF TEMPORARY
SERVICE POST OR SUPPORT. SUCH SUPPORT SHALL BE LOCATED
WITHIN 36" OF COMPANY'S TRANSFORMER OR SECONDARY VAULT,
OR FINAL METER LOCATION. CUSTOMER MUST PAY ANY APPLICABLE CHARGES.

2. ALL CUSTOMER WORK MUST BE COMPLETED AND REQUIRED INSPECTIONS
BE OBTAINED BEFORE COMPANY WILL PROVIDE SERVICE.

3. CUSTOMER WILL FURNISH AND INSTALL A NOMINAL 4" x 4" x 8' REDWOOD
OR OTHER CHEMICAL PRESERVATIVE TREATED POST OR OTHER COMPANY
APPROVED SUPPORT. A TEMPORARY OVERHEAD POLE OR SUPPORT CAN
BE USED PROVIDED APPLICANT REWIRES SUPPORT FOR UNDERGROUND
SERVICE.

4. INHIBITOR COMPOUND SHALL BE USED ON ALL ALUMINUM WIRE TERMINATIONS.

(8/73)

Fig. 9-15

Figures 9-16 through 9-18 show examples of various types of service-entrance equipment now in use. These should be compared to the drawings that follow in order to better visualize the riser diagrams such as Fig. 9-19 and plan views.

Fig. 9-16 Photograph of panelboard used in bank. See Fig. 9-20 for corresponding riser diagram.

Fig. 9-17

Fig. 9-18

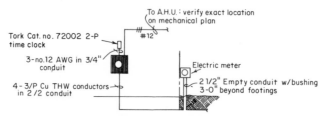

Fig. 9-19 Power-riser diagram.

PRACTICAL APPLICATIONS

The power-riser diagram shown in Fig. 9-20 was used on the plans for a branch bank, the floor plan of which was shown in

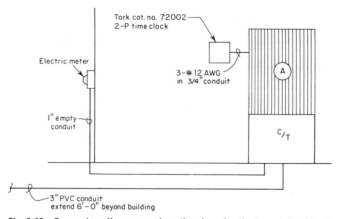

Fig. 9-20 Power-riser diagram used on the plans for the branch bank in Fig. 2-9.

Chapter 2. In analyzing this diagram, the following things are apparent:

1. A 3-in. PVC (plastic) conduit extends from 6 ft beyond the building to the C/T (current transformer) cabinet.

2. A 1-in. empty conduit runs from the C/T cabinet to the meter base mounted on the outside wall of the building. This conduit is for the power company's six meter wires from the current transformer.

3. Panel A is mounted above the C/T cabinet.

4. A two-pole time clock is mounted next to the panel and is fed with three No. 12 AWG conductors. This is to control the outside sign and building lights.

This power-riser diagram, in itself, is not sufficient to tell the electrical contractor exactly what is to be done. However, the written specifications along with the panelboard schedule (also on the drawings) will give all the necessary information. The panelboard schedule will give the manufacturer, voltage, amperage, and number and size of breakers required in panel A; the written specifications indicate that the local power company will pull the service lateral into the C/T cabinet which is also furnished and installed by the power company; the meter base, while furnished by the power company, is to be installed by the contractor along with the 1-in. empty conduit.

The power-riser diagram shown in Fig. 9-21 is very similar to

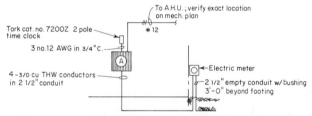

Fig. 9-21 A power-riser diagram similar to that shown in Fig. 9-19, except that more information is given directly on the riser diagram.

the diagram in Fig. 9-20 except that more information is given directly on the riser diagram. This riser diagram, when used in conjunction with the floor plans and the panelboard schedule shown in Fig. 9-22, eliminates the need for written specifications, provided that a note on the drawing states that "all work shall be done in accordance with the National Electrical Code."

Panel Board Schedule

Panel no.	Type Cabinet	Panel mains			Branches					Items fed
		Amperes	Volts	Phase	IP	2P	3P	Prot.	Frame	
"A"	Flush	200a.	120/240 V	3∅4WΔ	—	1	—	20 A	70 A	Time clock
Square "D" NQOB w/main breaker					—	—	1	20 A	70 A	A.H.U.
					—	1	—	30 A	70 A	Water htr.
					—	—	1	30 A	70 A	Cond. unit
					5	—	—	20 A	70 A	Lights
					10	—	—	20 A	70 A	Recepts
					5	—	—	20 A	70 A	Spares
					12	—	—	—	—	Provisions only

Fig. 9-22 Panelboard schedule from which the power-riser diagram in Fig. 9-21 was taken.

Each electrical system is required to be grounded in the manner prescribed by the National Electrical Code, although the details of grounding rarely appear in the power-riser diagram; rather, a paragraph usually appears in the written specifications and may read as follows:

> d. GROUNDING: The electrical contractor shall properly ground the electrical system as required by the National Electrical Code. The ground wire for the service entrance shall be run in conduit and connected to the main water service and connected ahead of any valve or cutoff.

The paragraph above will serve in most cases, but the electrical designer should carefully coordinate all electrical drawings with the plumbing drawings and the plumbing contractor because PVC (plastic) water pipe may be specified, and would not provide a ground.

It is recommended that the electrical designer calculate and give exact details of the grounding system for each service entrance. Such details may appear in the specification, as, for example, in the following specification for a 200-A service entrance:

> Furnish and install a No. 4 AWG bare copper conductor in a ³/₄-in. rigid galvanized conduit from the neutral bus in the main distribution panel to a 1-in. or larger cold-water pipe in the mechanical room on the ground floor. Water-pipe connections shall be with an approved ground fitting that bonds both conduit and conductor to water pipe.

The floor plan of an electrical equipment room for a department store is shown in Fig. 9-23. This room contains the main distribution panel, C/T cabinet, telephone cabinet, and other subpanels.

The corresponding power-riser diagram for this service entrance is shown in Fig. 9-24 and conveys the following information:

1. The service entrance consists of three 3½-in. conduits with four 500-MCM THW conductors in each, terminating in the C/T cabinet. This is rated for 1,200 A.

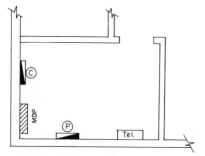

Fig. 9-23 Floor plan of an electrical equipment room.

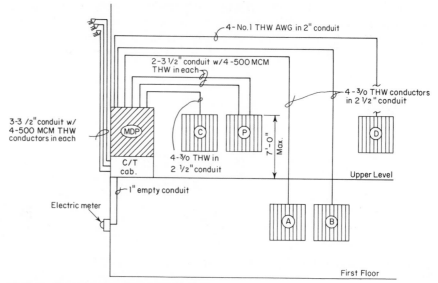

Fig. 9-24 The power-riser diagram for Fig. 9-23.

2. A 1-in. empty conduit runs from the C/T cabinet and terminates in a meter base mounted on the side of the building.

3. The main distribution panel is mounted over the C/T cabinet and is identified "MDP."

4. Panels A, B, C, D, and P are fed from the main distribution panel by feeders as indicated in the schedule. Details of each panel can be found in the panelboard schedules, elsewhere on the drawings.

The floor plan of a mechanical room for a three-story building is shown in Fig. 9-25. This room contains, besides mechanical

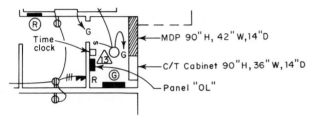

Fig. 9-25 Floor plan of a mechanical room.

equipment, the main distribution panel, C/T cabinet, emergency panel, panel P, panel OL, and a 100-A magnetic contactor for panel OL which is controlled by a time clock. The corresponding power-riser diagram is shown in Fig. 9-26; this shows

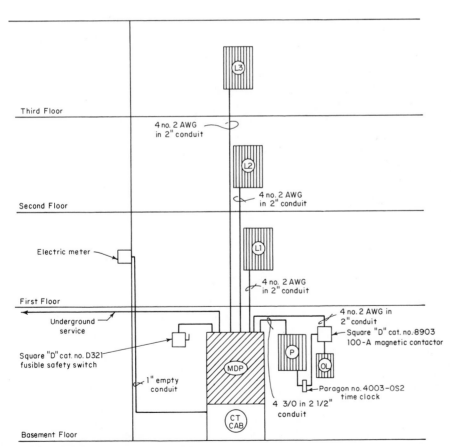

Fig. 9-26 Power-riser diagram for Fig. 9-25.

a schematic arrangement of all panels, feeders, and conduits as well as wire sizes for the power equipment illustrated.

RESIDENTIAL SERVICE-ENTRANCE EQUIPMENT

Several methods of sizing residential service entrances are given in Chap. 9, "Tables and Examples," of the National Electrical Code; one of them is very much as follows:

Example No. 1(b).   Single-Family Dwelling.

Optional Calculation for One-Family Dwelling (Section 220-7)

Dwelling has a floor area of 1,500 sq ft exclusive of unoccupied cellar, unfinished attic, and open porches. It has a 12-kW range, a 2.5-kW water heater, a 1.2-kW dishwasher, 9 kW of electric space heating installed in five rooms, a 4.5-kW clothes dryer, and a 6-A 230-V room air-conditioning unit.

$$\text{Air conditioner kW is } 6 \times 230/1,000 = 1.38 \text{ kW}$$

1.38 kW is less than the connected load of 9 kW of space heating; therefore, the air-conditioner load need not be included in the service calculation [see Section 220-4(k)].

1,500 sq ft at 3 W.	4.5 kW
Two 20-A appliance outlet circuits at 1,500 W each	3.0
Laundry circuit	1.5
Range (at nameplate rating)	12.0
Water heater	2.5
Dishwasher.	1.2
Space heating.	9.0
Clothes dryer	4.5
Total	38.2 kW

Calculations:

First 10 kW at 100 percent = 10.00 kW
Remainder at 40 percent (28.2 × 0.4) = 11.28 kW
Calculated load for service size is 21.28 kW or 21,280 W.

$$\frac{21,280}{230} = 92.5 \text{ A}$$

Therefore, this dwelling may be served by a 100-A service.

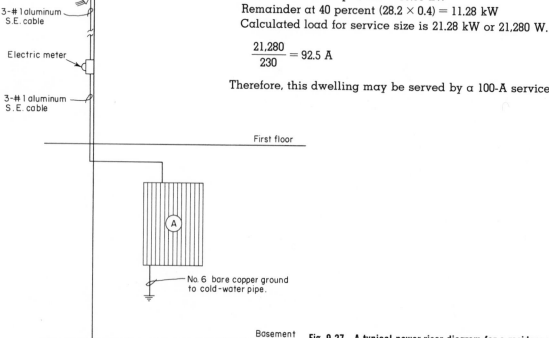

3-#1 aluminum
S.E. cable

Electric meter

3-#1 aluminum
S.E. cable

First floor

Ⓐ

No. 6 bare copper ground
to cold-water pipe.

Basement   **Fig. 9-27   A typical power-riser diagram for a residence.**

A typical power-riser diagram which may be used on the working drawings for this residence is shown in Fig. 9-27; a typical panelboard schedule is shown in Fig. 9-28.

PANELBOARD SCHEDULE										
Panel No.	Type Cabinet	Panel Mains			Branches					
		Amperes	Volts	Phase	IP	2P	3P	Prot	Frame	Item fed or remarks
A	Surface	100	120/240	IP, 3W	-	1	-	30A	70A	Clothes dryer
Square "D" type "QO" w/100A main					-	5	-	20A	70A	Space heating and water heater
					-	1	-	50A	70A	Range
					5	-	-	20A	70A	Appliances & ac unit
					6	-	-	15A	70A	Lighting & receptacles
					4	-	-	15A	70A	Spares

Fig. 9-28 A typical panelboard schedule for a residence.

HIGH-VOLTAGE SERVICE ENTRANCE

Figure 9-29 shows the site plan for a new building at a state institution. The existing transformer bank and primary service

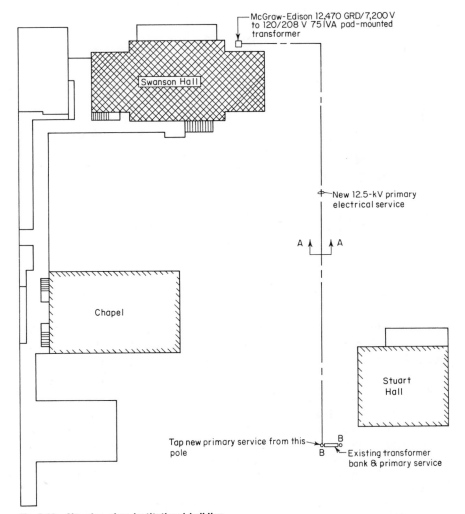

Fig. 9-29 Site plan of an institutional building.

are located in the lower right-hand corner of the drawing. The designer indicates that the new service is tapped at this point and runs underground to the new pad-mounted transformer.

Section A-A showing the detail of the buried cable is illustrated in Fig. 9-30.

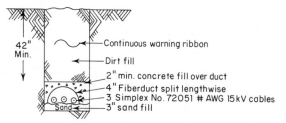

Fig. 9-30 Detail of the buried cable in the site plan in Fig. 9-29.

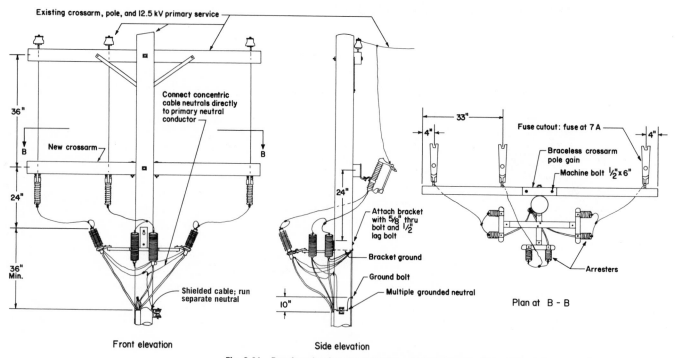

Fig. 9-31 Drawing showing the details involved in tapping the existing primary service.

The details involved in tapping the existing primary service are illustrated in Fig. 9-31. A note also appears on the drawings describing the underground primary cable. See Fig. 9-32.

NOTE: All high-voltage underground wiring shall be direct-burial, 15 kV, No. 4 alum. AWG, type AA, 2 conductor-1-No. 2 alum. AA, 7-strand phase conductor with extruded semiconducting cross-linked polyethylene strand shielding, 175 mils of crosslinked polyethylene insulation, 30 mils semiconducting polyethylene jacket and 10-No. 14 AWG bare copper concentric neutral; min. depth shall be 42″.

Fig. 9-32 Notes on drawings describing the underground primary cable.

A power-riser diagram showing the secondary service-entrance equipment is shown in Fig. 9-33.

The transformer-pad detail is shown in Fig. 9-34, and notes pertaining to the entire installation are shown in Fig. 9-35.

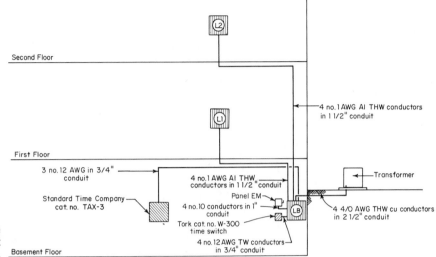

Fig. 9-33 Power-riser diagram showing the secondary-service–entrance equipment for this project.

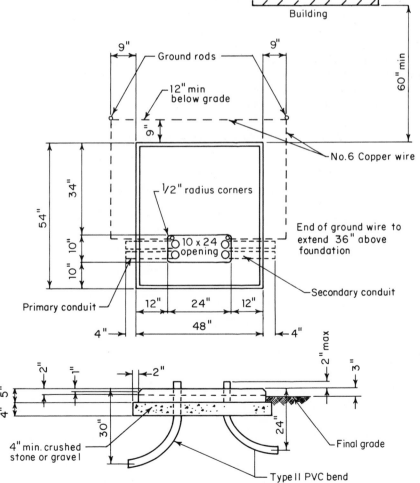

Fig. 9-34 Transformer-pad detail for the project shown in Fig. 9-29.

Underground Distribution Notes

1. To avoid physical obstructions and to provide adequate space separation for fire protection, the following are minimum clearances for locating transformer foundations: (For physical location see TD 2150-2)

    10 ft from window (along wall horizontally)

    10 ft below window (vertically)

    5 ft from building or other structure

    10 ft from door or entrance (along wall)

    10 ft from fire escape

    10 ft from ventilating ducts

2. Concrete pad may be poured in place or may be precast.

3. Install all conduits before placing pad. Conduits should not be placed under the section of the pad supporting the transformer so that the original ground will not be disturbed.

4. Conduit shall be rigid polyvinyl chloride.

5. Thoroughly compact the crushed stone or gravel.

6. To prevent water migration from concrete, place a waterproof membrane on crushed stone or gravel before placing concrete.

7. Backfill shall be clean granular soil, free of large stones and perishable material. All backfill shall be spread and compacted in maximum layers of 8 inches.

8. Where damage to the transformer by vehicles is possible, the transformer shall be protected by an appropriate barrier.

9. Seal all spare conduits and openings to prevent the entry of rodents and animals into the transformer compartment.

10. If conduit extends into building it must be sealed at the building end to prevent gas from entering the building through the conduit. Use Aqua Seal Stk. No. 594006.

11. Concrete to develop 3000 psi at 28 days' age, contain a minimum of 5.5 bags of cement per cu. yd. and a maximum of 6 gal. of water per 94-pound bag of cement, and conform to ASTM designation C-94.

12. Cement to be I of I-A and meeting ASTM designation C-150 and C-175, respectively.

13. Reinforcing steel shall conform to ASTM designation A615, Grade 60.

Fig. 9-35 Notes on the working drawings pertaining to the entire installation.

Chapter 10
Panelboard and Feeders

Branch-circuit control and overcurrent protection must be provided for all circuits and the power-consuming devices connected to them. Protective devices, such as fuses, circuit breakers, and relays, are usually grouped together at one centralized location (or more) in order to keep the length of the branch-circuit raceways at a practical minimum length for operating efficiency.

These protective devices are usually factory-assembled and installed in a metal housing, the entire assembly being commonly called a panelboard or "panel." The number and sizes of the overcurrent protective devices vary in accordance with requirements for lighting and power of the branch-circuit wiring system.

PANELBOARDS

On electrical drawings, a panelboard is usually indicated by a solid rectangle for surface-mounted panelboards such as

117

Fig. 10-1 Method of indicating a surface-mounted panelboard on working drawings.

Fig. 10-2 Method of indicating flush-mounted panelboard on working drawings.

shown in Fig. 10-1 and a solid rectangle with a "flange" for a flush-mounted panelboard as shown in Fig. 10-2. A mark is also used to identify each panelboard on the drawings: Fig. 10-1 is identified by a circle with the letter P, and Fig. 10-2 with the letter L.

The following material is a guide to various types of panelboards as furnished by one manufacturer.

Type CHP Panelboard

This type of panelboard utilizes the plug-in type of branch-circuit breaker for lighting and appliance branch circuits. Figure 10-3 shows typical wiring diagrams for both single- and three-phase panels, while Fig. 10-4 shows the approximate cabinet dimensions.

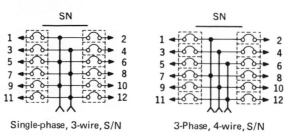

Single-phase, 3-wire, S/N   3-Phase, 4-wire, S/N

Fig. 10-3 Typical wiring diagrams for both single- and three-phase panels. (*By permission of Cutler-Hammer.*)

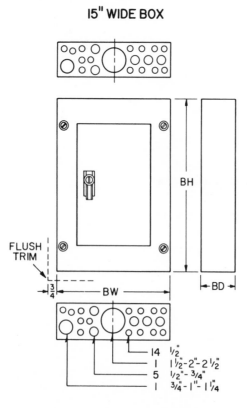

Gutters: Top and Bottom – 5" Min.
Sides – 4-1/2" Min.

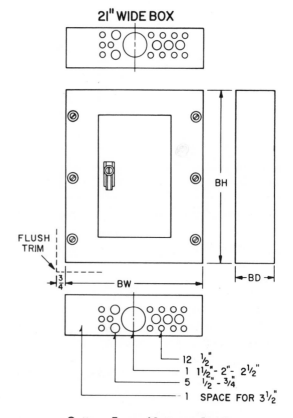

Gutters: Top and Bottom – 5" Min.
Sides – 7-1/2" Min.

Type CHB Panelboard

This type of panel is very similar to type CHP except that it is constructed for bolt-in-type circuit breakers for use on lighting and appliance branch circuits. The wiring diagram and physical dimensions shown for the CHP-type panel in Figs. 10-3 and 10-4 will also suffice for type CHB panelboards.

Type NA1B and NH1B Panelboards

This type of panelboard uses bolt-in-type EA-frame circuit breakers for lighting and appliance branch circuits. This type is designed for use with 277/480-V service and from 100- to 400-A capacity. Branch breakers are available as single-, two-, or three-pole types.

Type NDP and NHDP Panelboards

For use with 120/240-V or 120/208-V service using bolt-in-type EA-frame or EH-frame circuit breakers for power branch circuits. The NHDP panelboard is also available for use with 277/480-V service with the same types of frame. A suggested method of designing an electrical system for use with this type of panelboard is to make a sketch of the panelboard as shown in Fig. 10-5 (right).

List amperage capacity of all components, that is, neutral, main lugs, or current breaker and panel. If the total X-unit spaces required exceeds 33, lay out two or more panelboards. If these panels need to be cabled together, include subfeed lugs or circuit breaker.

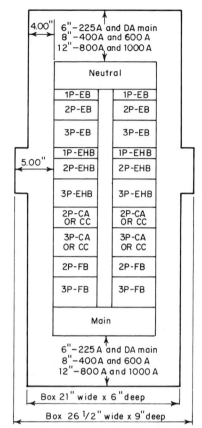

Fig. 10-5 Sketch used for designing NDP and NHDP panelboards.

Fig. 10-4 (On facing page and below) Approximate cabinet dimensions of single- and three-phase panels. (*By permission of Cutler-Hammer.*)

Number of poles	Main, A	Wide BW	High BH	Deep BD	Number of poles	Main, A and C.B. type	Wide BW	High BH	Deep BD
		Single-phase–three-wire							
8–20	125	15	19½	4⅛	8–10	50 EA	15	29	4⅛
20–30	225	15	29	4⅛	12–20	100 EA	15	29	4⅛
32–42	225	15	34	4⅛	20–30	225 QCC	15	38	4⅛
54	225	15	42½	4⅛	32–42	225 QCC	15	42½	4⅛
					20–42	225 JA	21	48¼	6
					54	225 QCC	21	54¼	6
					54	225 JA	21	58½	6
		Three-phase–four-wire							
8–18	125	15	19½	4⅛	8–12	50 EA	15	29	4⅛
20–30	125	15	29	4⅛	14–18	100 EA	15	29	4⅛
30–42	225	15	34	4⅛	20–30	100 EA	15	38	4⅛
54	225	15	42½	4⅛	30–42	225 QCC	15	42½	4⅛
					30–42	225 JA	21	48¼	6
					54	225 QCC	21	54¼	6
					54	225 JA	21	58½	6

The table is titled: **With Main Lugs Only** (left group) and **With Circuit-Breaker Main** (right group), with sub-columns Dimensions, inches (Wide BW, High BH, Deep BD).

NOTE: (Applies to panels standard in the 15″ wide box.) When conduit larger than 2½″ is to be used, or if more gutter space is desired, a 21″ wide by 6″ deep box of approximately the same height will be provided at no charge, upon request. Not available in unassembled panels.

5,000 AMP. INTERRUPTING CAPACITY SAFETYBREAKERS

Amps.	Catalog Number	Price	Packaged Quantities			Ship. Wt. Lbs. Each	Catalog Number	Price	Packaged Quantities			Ship. Wt. Lbs. Each
			Box	Carton	Package				Box	Carton	Package	
	SINGLE POLE—120/240 VOLTS A-C						DOUBLE POLE—COMMON TRIP—120/240 A-C					
15	CH115	$ 3.50	2	10	100	¼	CH215	$ 8.10	1	5	20	½
20	CH120	3.50	2	10	100	¼	CH220	8.10	1	5	20	½
20	❶CH120HM	3.50	2	10	100	¼	....	....	...	..	..	
25	CH125	3.50	2	10	40	¼	CH225	8.10	1	5	20	½
30	CH130	3.50	2	10	40	¼	CH230	8.10	1	5	20	½
40	CH140	3.50	2	10	40	¼	CH240	8.10	1	5	20	½
50	CH150	3.50	2	10	40	¼	CH250	8.10	1	5	20	½
60	....	....	...	..	..	..	CH260	8.10	1	5	20	½
70	....	....	...	..	..	..	CH270	16.40	1	5	20	½
❷90	....	....	...	..	..	..	CH290	22.00	1	..	10	1
❷100	....	....	...	..	..	..	CH2100	22.00	1	..	10	1
	DOUBLE POLE—COMMON TRIP—240 VOLTS A-C						THREE POLE—COMMON TRIP—240 V. A-C					
15	CH215H	$22.00	...	5	20	½	CH315	$27.60	...	..	4	¾
20	CH220H	22.00	...	5	20	½	CH320	27.60	...	..	4	¾
25	....	....	..	..	..		CH325	27.60	...	..	4	¾
30	CH230H	22.00	...	5	20	½	CH330	27.60	...	..	4	¾
40	CH240H	22.00	...	5	20	½	CH340	27.60	...	..	4	¾
50	CH250H	22.00	...	5	20	½	CH350	27.60	...	..	4	¾
60	CH260H	22.00	...	5	20	½	CH360	27.60	...	..	4	¾
70	CH270H	22.00	...	5	20	½	CH370	35.60	...	..	4	¾
❷90	CH2090H	25.60	...	..	10	1	CH3090	40.70	1	..	5	1½
❷100	CH2100H	25.60	...	..	10	1	CH3100	40.70	1	..	5	1½
	2 WIRE SWITCHED NEUTRAL 120/240 VOLTS A-C						3 WIRE SWITCHED NEUTRAL 120/240 VOLTS A-C					
15	CH215SW	$10.90	...	..	4	½	....	....	...	..	..	
20	CH220SW	10.90	...	..	4	½	CH320SW	$ 16.40	...	..	4	¾

THREE POLE — DELTA — 240 VOLTS A-C

Amps.	Catalog Number	Price	Quantity per Package	Ship. Wt. Lbs. Each	Amps.	Catalog Number	Price	Quantity per Package	Ship. Wt. Lbs. Each
20	CH320D	$27.60	1	¾	50	CH350D	$27.60	1	¾
30	CH330D	27.60	1	¾	60	CH360D	35.60	1	¾
40	CH340D	27.60	1	¾	...	......	....	....	..

DOUBLE POLE FOR WATER HEATERS — COMMON TRIP — 120/240 VOLTS A-C

Amps.	Catalog Number	Price	Qty per Package	Ship. Wt.	Amps.	Catalog Number	Price	Qty per Package	Ship. Wt.
15	CH215WH	$11.50	4	¾	30	CH230WH	$11.50	4	¾
20	CH220WH	11.50	4	¾	...	......	....	....	..

❶ High magnetic trip.
❷ 90 and 100 ampere require 2 spaces per pole.

Fig. 10-6 Table of circuit breakers available.

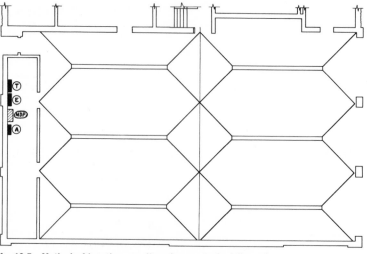

Fig. 10-7 Method of locating panelboards on a typical floor plan.

The labeled sketch should then be used for a neat drawing on the working drawings of the project, for example, in the power-riser diagram.

Type CDP Panelboard

This type of panelboard also uses bolt-in circuit breakers for power branch circuits and feeder distribution in ratings from 15 through 1,200 A. The maximum voltage rating for dc power is 250 V; ac power rates up to 600 V. The table in Fig. 10-6 gives the various circuit breakers available together with related information.

METHOD OF SHOWING PANELBOARDS
ON WORKING DRAWINGS

Panelboards usually are shown on all floor plans of buildings as well as in a power-riser diagram. The panels are first located on the floor plan as shown in Fig. 10-7. This enables the estimator to accurately calculate the length of the panel feeders as well as the branch circuits to them. Each panel (as can be seen in Fig. 10-7) is given an identifying letter or number to which reference can be made.

A panelboard schedule should be provided on the working drawings or in the specifications to relay the following information:

1. Panel identification
2. Type cabinet (surface- or flush-mounted)
3. Panel mains
   a. Amperage rating
   b. Voltage
   c. Phase
4. Number of circuit breakers including number of poles, trip amperage, frame type, etc.
5. Items fed or remarks

A typical panelboard schedule is shown in Fig. 10-8.

Panelboard Schedule									
Panel No.	Type Cabinet	Panel Mains			Branches				Item Fed or Remarks
		Amps	Volts	Phase	1P	2P	3P	Prot	Frame
"C"	Surface	100a	120/240	3plu, 3W	–	–	2	20A	70A	Printing equipment
Square "D" type NQO					–	1	–	50A	70A	Furnace
W/M.L.O.					10					Provisions only

Fig. 10-8 Typical panelboard schedule.

Most working drawings also include a power-riser diagram to indicate how each panel is fed, that is, the number and size of conductors and the size of the conduit enclosing the conductors. A typical power-riser diagram is shown in Fig. 10-9.

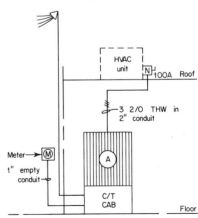

Fig. 10-9 Typical power-riser diagram.

FEEDER AND PLUG-IN DUCT

The modern approach to flexible power distribution in industrial electrical systems tends to carry feeder conductors in troughs or trays or to use bus bars enclosed in either a ventilated or a nonventilated duct from the source of supply to points within the system where distribution of electric power is needed.

Bus ducts provide for fast initial installation and are also very fast and convenient for adding on electrical feeders at a later date. Straight ducts are manufactured in 1½-, 3-, 5-, 6-, and 10-ft lengths. These are bolted together by means of bolt bus bar joints. Numerous fittings are also available as can be seen in Fig. 10-10.

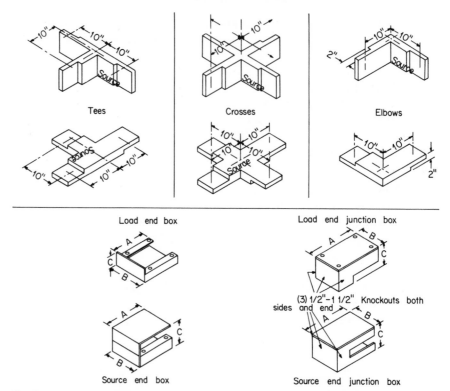

Fig. 10-10 Numerous fittings are available for use with feeder duct. (*By permission of Cutler-Hammer.*)

Fitting	Number of poles								
	3-, 4-, and 5-Pole			6- and 7-Pole			8- through 13-Pole		
	A, in.	B, in.	C, in.	A, in.	B, in.	C, in.	A, in.	B, in.	C, in.
Load end box	6¼	6¹/₁₆	2	10¾	7⁹/₁₆	7	18	7⁹/₁₆	7
Source end box	6¼	6¹/₁₆	2	10¾	7⁹/₁₆	7	18	7⁹/₁₆	7
Load end junction box	10¾	6¹/₁₆	4	10¾	7⁹/₁₆	7	18	7⁹/₁₆	7
Source end junction box	10¾	6¹/₁₆	4	10¾	7⁹/₁₆	7	18	7⁹/₁₆	7

The broad array of feeder and plug-in duct types and standardized fittings will satisfy almost any low-voltage requirement for local circuit protection and control. Plug-in devices—fused, fused disconnect switches, circuit breakers,

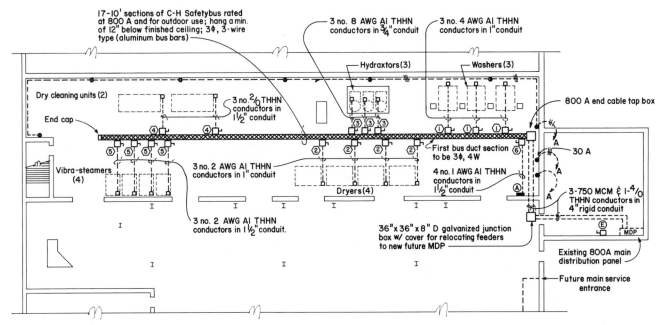

Fig. 10-11 Floor plan of a commercial laundry room.

transformers, etc.—are readily available with feature rating and characteristics to match nearly any load condition.

A carefully planned network of bus ducts is necessary to provide maximum economy, flexibility, and low voltage-drop characteristics. Also because of physical limitations and electrical requirements, each design of a bus-duct system requires a customized solution. The following project will show the reader how one simple bus-duct system was designed and will apply—basically—to most bus-duct systems.

PROBLEM

A uniform-rental service required that a laundry room be supplied with electrical power to operate sixteen 240-V machines, a reasonable number of convenience outlets, and lighting for this area as well as for a small office space above. The basic floor plan is shown in Fig. 10-11.

1. Since branch-circuit wiring was the major item, the designer determined that in no instance should a machine or other load be more than 30 ft away from the main bus duct. Therefore, an end cable-tap box was located at one end of the area which was closest to the main power source. The total demand for this area was calculated to be 800 A. Therefore, two 4-in. conduits were run from the main power source to the end cable-tap box, each containing three 750-MCM THHN cables and one 4/0 aluminum conductor.

2. Eight 10-ft sections of bus duct were then specified to be attached to the end cable-tap box and to run along the approximate center of the area. Since taps were required all along this bus duct, the plug-in type was used throughout. However, in instances where a sizable run is required before taps are necessary, feeder-type ducts should be used from the power source out to the area where plug-in (branch-circuit) convenience is

required. This is because the price-per-foot cost of feeder duct is less than the plug-in type.

3. Note that, in the drawing in Fig. 10-11, the bus duct runs over aisles. This provides easy access to the duct and reduces installation costs as well as maintenance expense.

4. It was determined that the 800-A system would be the most satisfactory and most practical for other areas within the plant; if one ampere size is standard throughout the plant, any later modification of the system can be readily made without a lot of additional expense. If additional ducts need to be added or if part of the system must be disassembled and moved to a new location, all components can be reapplied with very little planning.

5. Although this is not required in the project under consideration, if extensive runs are needed, it might be advisable to install more runs on closer centers to obtain better blanket coverage and reduce the branch-circuit wiring costs.

6. Since there are more than six distribution circuits in this system (not shown in Fig. 10-11), a main was required on the main distribution panel. However, where possible, try to arrange a system so that there are no more than six distribution circuits. With this consideration, according to the National Electrical Code no main breaker or service switch is required.

7. Notice that lighting panel A is also fed from the bus duct, eliminating the cost of an additional feeder from the main distribution center. Always check the advantages of connecting lighting in with the bus-feeder system. It is usually more economical to install a four-wire system and tap the lighting panels directly off the bus duct. In our example, the main service is 120/240 V, so no further modification was necessary for the lighting. When a 480-V service is required for the main bus duct, dry-type transformers are available which attach directly to the bus duct for 120-V loads.

8. For greater installation economy, the bus duct in our example was mounted edgewise because installation procedures are quicker, the edgewise hangers are more economical, and the costs in general are lower.

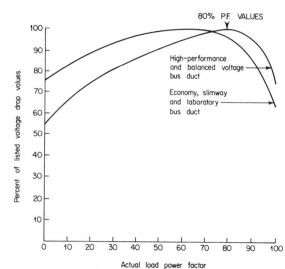

Fig. 10-12 Graph showing voltage-drop and power-factor figures.

Switch Schedule									
Switch Type	Manufacturer's Description	Volts	Amperes	Prot.	Conduit Size	Conductors		Remarks
						No.	Size	
①	Cutler–Hammer cat. no. CP2HD 323	240	100	60A	1"	3	No. 4	Verify exact connection details
②	cat. no. CP2HD 323		100	90A	1"	3	No. 2	
③	cat. no. CP2HD 32		30	25A	3/4"	3	No. 8	
④	cat. no. CP2HD 324		200	125A	1 1/2"	3	2/0	
⑤	cat. no. CP2HD 323		100	90A	1 1/2"	3	2/0	
⑥	cat. no. CP2HD 423		100	100A	1 1/2"	4	No. 1	

Fig. 10-13   Schedule to identify switches used on bus duct.

PLANNING THE SYSTEM

A duct system can be planned easily and quickly in most cases. The basic factors necessary to electrically plan and execute a compatible, efficient, and economical system include:

1. Determine connected-load and future-growth requirements.
2. Determine the load demand and duty cycles of the various loads.
3. Calculate voltage drop.
4. Calculate power factors.

Factors 1 and 2 are acquired from the owners or equipment suppliers.

The voltage drop can be calculated by the following formula:

$$\text{Maximum voltage drop per 100 ft} \times \frac{\text{actual load amperage}}{\text{rated amperage}}$$
$$\times \frac{\text{percentage drop}}{\text{@ worst power factor}} = \text{voltage drop per 100 ft}$$

The graph in Fig. 10-12 will aid the designer in voltage-drop and power-factor calculations.

EXAMPLE

A copper bus duct rated at 800 A is installed with a load of 400 A (demand) at 0.60 power factor. Voltage drop (2.8) × 400 load A/800 rated A × percent of voltage drop at appropriate power factor (0.93) = 1.3 V per 100 ft. Therefore the complete voltage drop over a 500-ft run would be 6.5 V.

In our example in Fig. 10-11, plug-in taps are required adjacent to each piece of machinery and at the lighting panel; these switches are indicated by the symbol ⬡ with an identification number at each. A schedule is then shown on the working drawing to exactly identify each switch. This schedule is shown in Fig. 10-13.

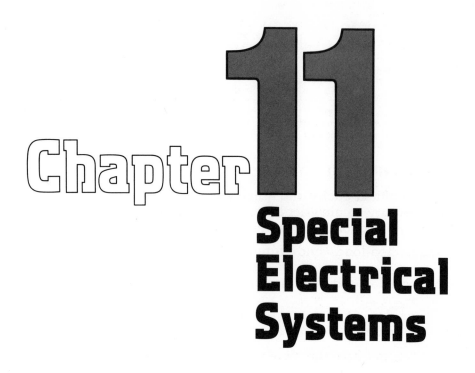

Chapter 11
Special Electrical Systems

REMOTE-CONTROL SYSTEMS FOR LIGHTING CIRCUITS
Remote-Control Relay Switching

Relays were developed several years ago to meet the demands for more convenient and reliable control of industrial, commercial, and residential equipment. The relays permitted more flexible control for residential lighting, and low voltage and low current in the control circuit provided for added safety.

The basic circuit of a remote-control wiring system is as shown in Fig. 11-1.

The split-coil type of relay, as shown in Fig. 11-1, permits positive control for ON and OFF. The relay can be located near the lighting load, or installed in centrally located distribution panel boxes, depending on the application.

Because no power flows through the control circuits and low voltage is used for all switch and relay wiring, it is possible to

Relay

On
Off
Blue

Red
White
Black

On
Off

Switch

Relay

Load

Branch
circuit

24V
Transformer

Line

Fig. 11-1 Basic circuit of a remote-control wiring system.

Fig. 11-2 Typical 24-V transformer used for the power supply in remote-control systems.

place the controls at a great distance from the source or load, thus offering many advantages through this modern system of wiring.

System Components

Power Source. A single-main power source is used to provide the correct voltage for operation of all relays in a remote-control system. The power supply usually consists of a pulse-type transformer, a selenium rectifier, and an electrolytic capacitor. Input voltage is 120 or 277 V while the momentary output voltage occurring when a switch is pressed is 24 V. This power supply is usually mounted on one of the relay cabinets, but it may be installed at other locations, such as at the service entrance, if required for a particular application. Figure 11-2 shows a typical transformer used for the power supply in remote-control systems.

Relay. The relay normally used in residential remote-control systems is a mechanical latching type with a single coil. Momentary impulses from the power supply alternately open and close the contacts in the load circuit. Each relay may control several outlets if desired, but a single switch may not control more than one relay. See Fig. 11-3.

Switches. One basic type of switch, the single-pole, double-throw, momentary-contact push button, does the work of single-pole, double-pole, three-way, and four-way switches used in conventional wiring—and with less wiring. Pressing the ON position of the switch energizes the related relay and closes the circuit to the load; pressing the OFF position of the switch again energizes the relay and opens the circuit to the load. One type of remote-control switch does not have "on-off" designations, but with each touch of the push button, the relay alternately opens and closes. Figure 11-4 shows one type of remote-control switch.

Fig. 11-3 Remote-control relay.

Fig. 11-4 Remote-control switch.

Figure 11-5, which was supplied by Electro Systems, Inc., shows a typical relay cabinet, local switch connections, and related wiring.

Fig. 11-5 Typical relay cabinet.

Master-Selector Switch. Where many circuits must be controlled from one convenient location, master-selector switches as shown in Fig. 11-6 perform the necessary functions of selecting only those circuits wanted.

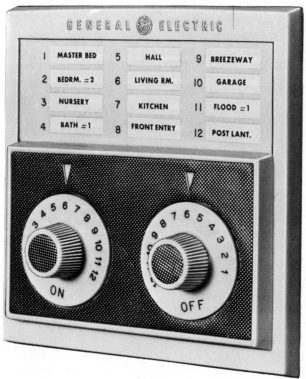

Fig. 11-6 Master-selector switch.

A "dial-type" master-selector switch for 12 circuits permits individual control of the circuits; alternatively, by pressing and then sweeping all circuits, master control is accomplished.

Motor-Master Control Units. Motor-driven master-control units are used to turn on or off up to 25 individual circuits; when these units are cascaded, the pressing of a single master switch can control any number of circuits, depending on the number of Motor-Master units (Fig. 11-7) that are ganged together. Ganging is accomplished by having the last position of the first unit wired to start the second unit, the last position on the second unit wired to start the third unit, and so on. These units can be activated by photoelectric cell relays, for automated lighting control.

Combining Remote Controls with Dimmers. Sometimes it is desirable to dim certain lighting fixtures which are controlled by a remote-control system. This is accomplished by wiring a dimmer between the relay and the fixture, as shown in the wiring diagram in Fig. 11-8. The dimmer adjusts the intensity of the light, and the remote controls turn it on or off.

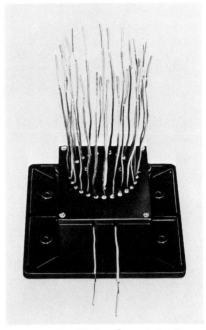

Fig. 11-7 Motor-Master ®️ control unit.

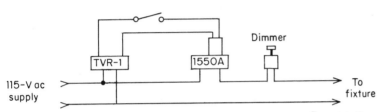

Fig. 11-8 Control-circuit connections used for combining remote controls with dimmers.

LOW-VOLTAGE CONTROLS

Low-voltage controls offer a safe, quiet means of switching electricity that is not possible to obtain with conventional switching methods, and they have provided a new freedom in the design and use of circuit control.

With a low-voltage control system, it is practical for the designer to provide—at a nominal cost—sufficient multiple-switch control of individual electrical circuits to create a new level in convenience for the user. Master-control centers also provide for remote control in the operation of multiple circuits from one or more locations and give visual indication of the use of each circuit. This capability in controlling lighting systems allows the electrical designer unusual latitude in providing his or her clients with the convenience desired.

The following discussion indicates several applications possible with low-voltage control and was supplied by Touch-Plate, Electro-Systems, Inc.

Visual Aids for Schools

A variety of visual aids, such as blackboards, flip charts, maps, etc., can be illuminated by supplemental spot and flood lamps, each connected to a master-control center at the instructor's podium. Lights can be controlled, as required, during instruction so that the desired visual aid can be highlighted. A wiring diagram for this application is shown in Fig. 11-9.

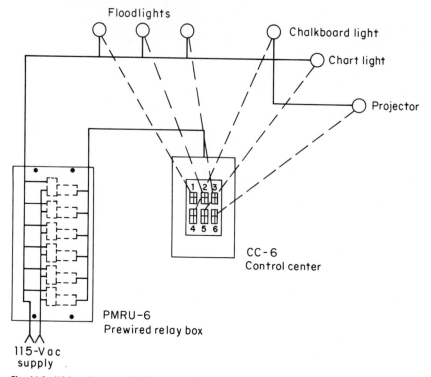

Fig. 11-9 Wiring diagram showing connections when a master-control center is used in conjunction with low-voltage relays for visual aids in schools.

Fail-Safe Alarm System

The purpose of this type of alarm is to alert the bank manager or security office when any one of the teller stations is in need of assistance, without sounding a general alarm.

Figure 11-10 shows the wiring diagram for a possible solution. Momentary contact switches are installed at each teller station and are wired with low-voltage wire to control centers in the offices of the manager and/or security officer. The control centers show at a glance the source of the warning. A bell or buzzer may also be installed which can be turned off by key as action is taken.

Industrial Application

PROBLEM

Lack of switch points owing to high ceilings and open truss construction. Attendant made rounds, turning out unused but sometimes necessary lights.

SOLUTION

A low-voltage remote-control system was installed. Number 18 AWG wire was strung along girders, reaching previously inaccessible switching points. Larger areas could now have zone control. Control panels at main entries revealed which

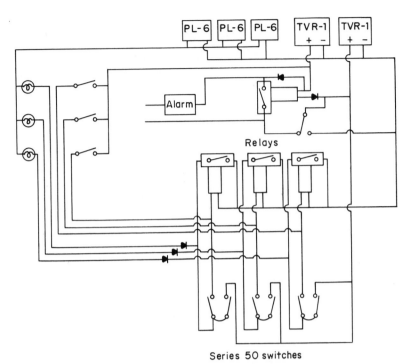

Fig. 11-10 Example of remote-control circuits used for security purposes.

lights were burning and gave fingertip control, thus reducing operating cost. See Fig. 11-11 for the wiring diagram.

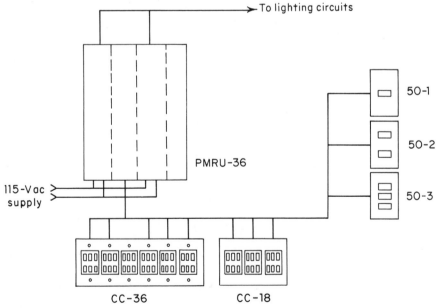

Fig. 11-11 Industrial application of remote-control circuits.

Motel Management

A low-voltage master-control center installed in a motel office and also in the supply room can save much time in determining which rooms are ready for occupancy. (See Fig. 11-12.) As soon as maid service is completed, the office is signaled from the supply room. When guests check out, the manager signals for maid service, maintaining a constant readiness and flow of work.

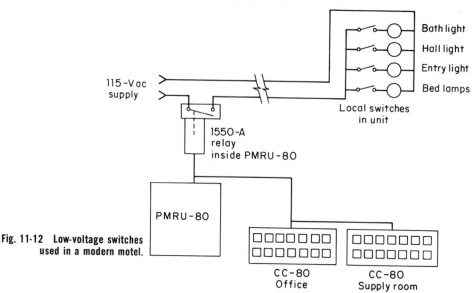

Fig. 11-12 Low-voltage switches used in a modern motel.

Home Protection

Floodlights are installed at the eaves of a house. These, along with entrance lighting fixtures and three landscape lighting circuits, were connected to a master-control panel in the master bedroom. When there are sounds or visitors in the night, occupants can instantly light up the entire property. See Fig. 11-13.

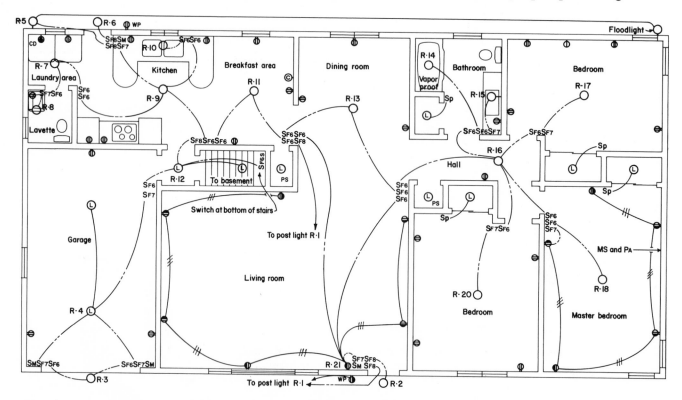

Nurse Call System

When it is necessary to alert the nursing-station supervisor that a patient has called for assistance, a low-voltage switch at the patient's bed can be used to turn on a corridor light outside the room as well as a light in the master-control center at the duty station. If it is wired to a wall switch at the door in the patient's room, the light at the control center will remain on until the patient receives attention. Then the light can be turned off by the wall switch in the patient's room. Figure 11-14 shows a wiring diagram of such a system.

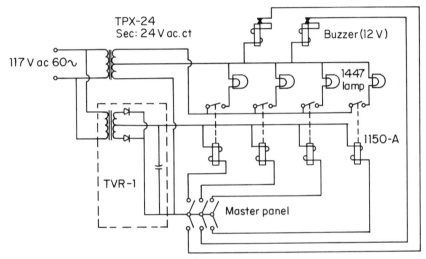

Fig. 11-14 Remote-control circuits used with a nurse-call system.

Nursing Homes

Low-voltage switches can provide ample, easy-to-handle switching for lights in nursing homes so that elderly patients need not walk in darkness or risk shock from wiring. In addition, master-control centers can provide zone lighting, allowing entire areas to be lit at once in case of emergencies. See Fig. 11-15.

Travel Trailers and Yachts

When space limitations are a problem, low-voltage switching is very practical, since the 18-gauge wire requires a minimum of space in the raceways and only a 1/2-in. depth is necessary for the switches. Because switch legs carry low voltage, danger of

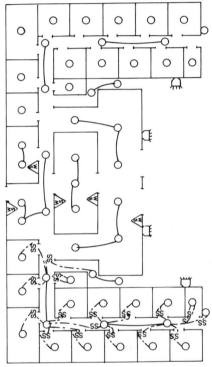

Fig. 11-15 Application of the remote-control circuit used in a nursing home.

Fig. 11-13 (Facing page) Remote-control circuits used for home protection. This layout shows a deluxe wiring system employing the use of multipoint switching that permits turning on room lights from every doorway, switch-controlled split receptacles in living room and master bedroom, master-selector switch at the bedside. Selected lights and receptacles are also controlled from two Motor-Master units not shown. It has been designed to provide protection control, turning on perimeter lights to floodlight the grounds, as well as lighting selected rooms from the master bedside. Pressing a single button S_M at any entrance, or the 12th position of the master-selector switch, starts the Motor-Master for ON control of the selected lights to provide a pathway of light through the house, or the OFF side for turning off the lights in and around the house. And here is downright convenience, a switch-controlled outdoor weatherproof split receptacle that lets the homeowner turn the decorative Christmas lights on and off from two handy indoor locations. Any portion of this layout can be used as a guide in laying out a General Electric remote-control installation to meet requirements specified by the architect, builder, or home-buyer. (NOTE: Be sure to specify pilot-type relays for pilot-light switches.)

fire and personal injury is kept to a minimum. Figure 11-16 shows a wiring diagram for such a system.

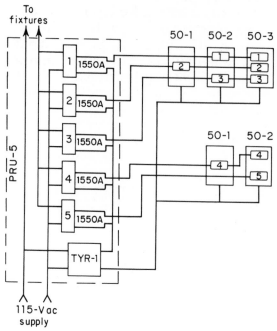

Fig. 11-16 Use of remote control in travel trailers and yachts.

Office Use

Should a large firm need to know where its key personnel are at all times, low-voltage master-control panels installed in each office and activity area can be wired into a similar panel at the receptionist's desk. By assigning a color to each person and a number to each room, it is possible to tell at a glance where each person is at any given moment. It is also possible to know who is in a particular office. See Fig. 11-17.

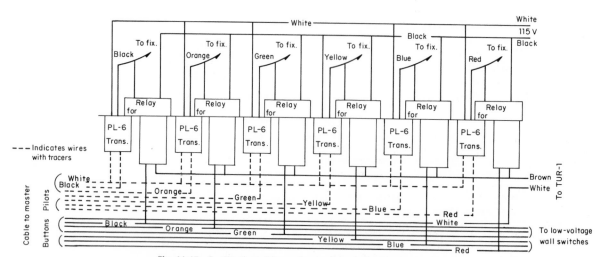

Fig. 11-17 Application of low-voltage controls in office areas to locate personnel.

Modular Housing

PROBLEM

Exterior walls were precast. Thin, interior wall-partitions were designed to be rearranged as changing space needs dictated.

SOLUTION

Conduit and outlet boxes were installed in the precast sections. All interior sections were wired with double runs of low-voltage wire so that low-voltage switches could be installed at either side of the panel, however it was placed. Multibutton master-control centers provided maximum control of all circuits. See Fig. 11-18.

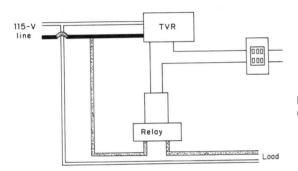

Fig. 11-18 Use of low-voltage controls in modular housing.

Restaurants

Sometimes it is desirable for restaurant managers to know instantly when a specific table anywhere in the restaurant becomes available for seating, without a lot of leg work.

A lighted scale model of the restaurant floor plan could be wired in parallel with master-control centers at two serving stations. The moment a table is cleared, the waiter pushes a button on his panel, signaling the manager that the table is ready. See Fig. 11-19.

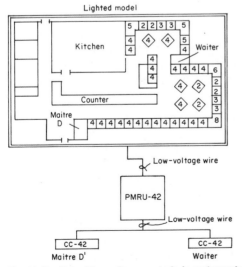

Fig. 11-19 Use of low-voltage controls in restaurants.

Figure 11-20 shows the floor plan of a typical residence using low-voltage switching for all lighting control.

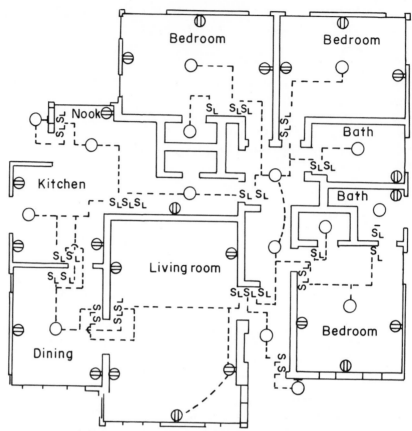

Fig. 11-20 Floor plan of residence using low-voltage switching for all lighting controls.

Figure 11-21 illustrates a typical relay cabinet and local switch connections for a typical low-voltage control system.

SIGNAL AND COMMUNICATION SYSTEMS

The field of electric signaling and communications is a very broad one, covering everything from simple residential door chimes to elaborate building-alarm and detector systems. For this reason, this chapter will not deal at length with any one branch of this field; rather it will touch briefly upon practically all branches with which the electrical designer will come in contact.

Electric Chime Systems

One of the simplest and most common electric signal systems is the residential door-chime system. Such a system contains a low-voltage source, a push button, wire, and set of chimes. The quality of the chimes will range from a one-note device to those which "play" lengthy melodies.

The wiring diagram in Fig. 11-22 shows a typical two-note chime controlled at two locations. One button, at the main entrance, will sound the two notes when pushed, while the other button, at the rear door, will sound only one note when pushed.

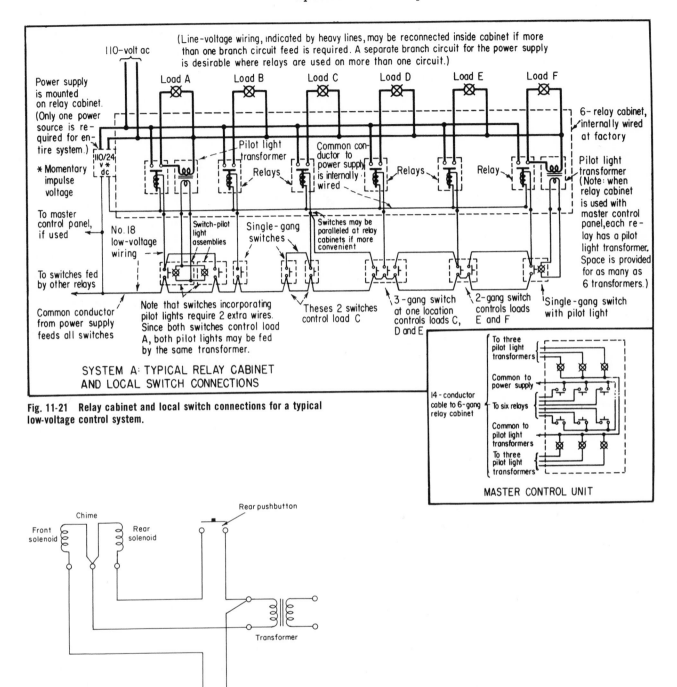

Fig. 11-21   Relay cabinet and local switch connections for a typical low-voltage control system.

Fig. 11-22   Wiring diagram showing a typical two-note chime control at two locations.

Fire-Alarm Systems

All fire-alarm systems fall into four basic types:

1. Noncoded
2. Master-coded
3. Selective-coded
4. Dual-coded

Each of these four types has several functional features, so that an electrical designer can design a specific system to the client's needs within local and state fire codes, statutes, and regulations.

In a noncoded system, an alarm signal is sounded continuously until manually or automatically turned off.

In a master-coded system, a common-coded alarm signal is sounded for not less than three rounds. The same code is sounded regardless of the alarm-initiating device activated.

In a selective-coded system, a unique coded alarm is sounded for each firebox or fire zone on the protected premises.

In a dual-coded system, a unique coded alarm is sounded for each firebox or fire zone to notify the owner's personnel of the location of the fire, while noncoded or common-coded alarm signals are sounded separately to notify other occupants to evacuate the building.

The most difficult aspect of applying fire-alarm systems is the selection of the basic system along with the functional features required by different occupancies. The following chart simplifies this selection process, for the chart reads easily yet covers 64 standard systems.

GUIDE CHART FOR FIRE-ALARM SYSTEMS

Type of system	General alarm annunciation			Presignal annunciation		
	None	"A" type	"D" type	None	"A" type	"D" type
Noncoded systems	N-1	NA-1	ND-1	NM-1	NAM-1	NDM-1
	N-2	NA-2	ND-2	NM-2	NAM-2	NDM-2
Common-coded systems	C-1	CA-1	CD-1	CM-1	CAM-1	CDM-1
	C-2	CA-2	CD-2	CM-2	CAM-2	CDM-2
Selective-coded systems:						
Series noninterfering,	S-1	—	—	SM-1	—	—
successive	S-2	—	—	SM-2	—	—
Positive noninterfering,	P-1	—	PD-1	PM-1	—	PDM-1
successive	P-2	—	PD-2	PM-2	—	PDM-2
Zone-coded	—	ZA-1	ZD-1	—	ZAM-1	ZDM-1
	—	ZA-2	ZD-2	—	ZAM-2	ZDM-2
Dual-coded systems:						
Noncode + series non-	NS-1	—	—	—	—	—
interfering, successive	NS-2	—	—	—	—	—
Noncode + positive non-	NP-1	—	NPD-1	—	—	—
interfering, successive	NP-2	—	NPD-2	—	—	—
Noncoded + zone-coded	—	NZA-1	NZD-1	—	—	—
	—	NZA-2	NZD-2	—	—	—
Common-coded + series	CS-1	—	—	—	—	—
noninterfering, successive	CS-2	—	—	—	—	—
Common-coded + positive	CP-1	—	CPD-1	—	—	—
noninterfering, successive	CP-2	—	CPD-2	—	—	—
Common-coded + zone	—	CZA-1	CZD-1	—	—	—
coded	—	CZA-2	CZD-2	—	—	—

The system numbers used in the preceding table consist of up to three alphabetical characters followed by the suffix 1 or 2.

The alphabetical characters denote:

N. A no-code system.

M. A multialarm or presignal system where the initial operation of any manual or automatic alarm-initiating device sounds an alarm on certain primary alarm signals. Authorized personnel with special keys may initiate a general-evacuation alarm from any manual firebox. On automatic fire-alarm systems, auxiliarized city fireboxes (if used) are tripped on the first alarm, and general-evacuation alarms are all noncoded.

C. A common-code or master-code system.

S. A selective-code system using coded fireboxes in which all fireboxes are in one common wired loop, and the coded firebox nearest the control unit has precedence on the loop. This is commonly called series noninterfering, and successive.

P. A selective-code system using coded fireboxes in which all fireboxes are on one common wired loop and arranged so that only one code at a time can sound even if two or more fireboxes are activated simultaneously. This prevents confusion resulting from incomplete, lost, or mutilated codes. This is commonly referred to as positive, noninterfering, and successive.

Z. A selective-code system using noncoded fireboxes or automatic devices that are point-wired to the control unit and with the coding done with the control unit. All codes sound in a positive, noninterfering, and successive manner as described for the P systems. All Z systems are furnished with annunciators in the control unit.

A. An annunciator feature on some systems in which each alarm zone illuminates a backlighted window. In addition, each alarm-zone window has an adjacent trouble window that lights in the event of an "open" condition on its associated alarm-zone wiring. Two lamps are used in the alarm window and one lamp in the trouble window. A flashing alarm-lamp feature is available with the A-line annunciators.

D. An annunciator feature on some systems in which each alarm zone illuminates a backlighted window. Two lamps are used in parallel so that the window can be read even if one lamp is burned out.

The suffixes 1 and 2 have to do with the alarm-signal circuits:

1 Alarm-signal operating voltages derived from the primary power source of 120 V ac

2 Alarm-signal voltage of 24 V dc derived from a transformer and rectifier isolated from the 120 V ac

Individual system descriptions are available from the manufacturer.

Fire-Alarm System Applications

Nursing Home. Figure 11-23 shows a partial floor plan of a nursing home. This portion of the floor plan shows the fire-alarm panel, smoke detectors (SD), striking stations, gongs

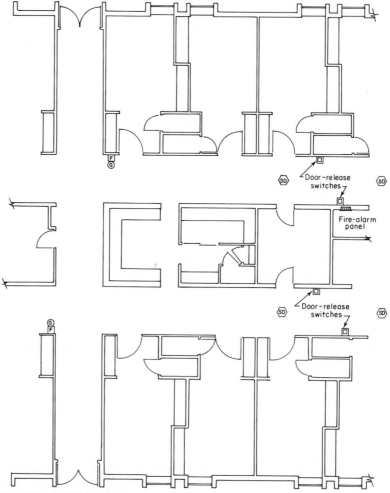

Fig. 11-23 Partial floor plan of nursing home.

(bells), and magnetic door-release switches. The fire-alarm riser diagram in Fig. 11-24 shows all devices connected to this system, along with the wiring of each.

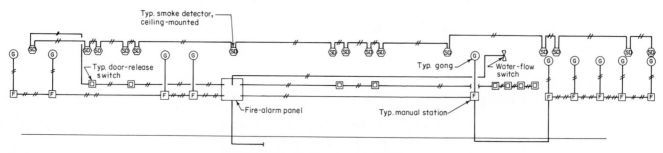

Fig. 11-24 Fire-alarm system riser diagram.

Basically, if any smoke detector senses smoke, or if any manual striking station is operated, all bells within the building will ring, indicating a fire. At the same time, the magnetic door switches will release the smoke doors to help block smoke and/or drafts. This system is also connected to a water-flow switch on the sprinkler system, so that if any of the sprinkler valves are activated, causing a flow of water in the system, the fire-alarm system will again go into operation energizing all bells and closing smoke doors.

Public Building.   The floor plan of a county courthouse is shown in Fig. 11-25. The fire-alarm panel is located in the mechanical room on another floor, but all striking stations, bells, door-

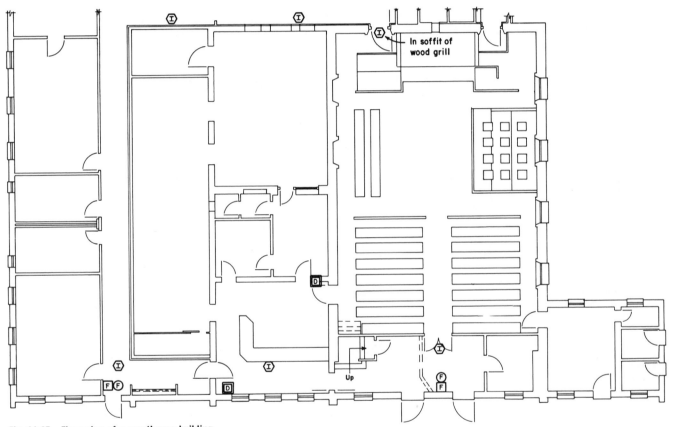

Fig. 11-25   Floor plan of a courthouse building.

release switches, and ionization smoke detectors are shown on this plan and operate in the same manner as the system described for the nursing home. A few differences, however, should be noted and can be seen by referring to the related fire-alarm riser diagram in Fig. 11-26.

1.   There is no sprinkler system in this building, and therefore no water-flow switch was used.

2.   A magnetic contactor was supplied on each of the three air-handling units used for air-conditioning purposes. When any part of the fire-alarm system is activated, either manually or automatically, all fans are automatically shut down to stop the flow of air within the building. This action keeps the spread of fire to a minimum.

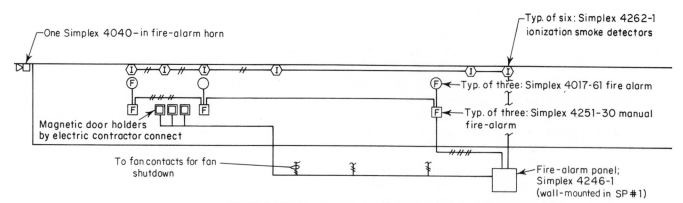

Fig. 11-26 Fire-alarm system riser diagram used for the courthouse drawings.

3. The ionization smoke detectors are more sensitive than the conventional smoke detectors used on the nursing home project.

INTERCOM SYSTEMS

A local rescue squad unit recently built an addition onto their existing building. During the planning stages of this addition, it

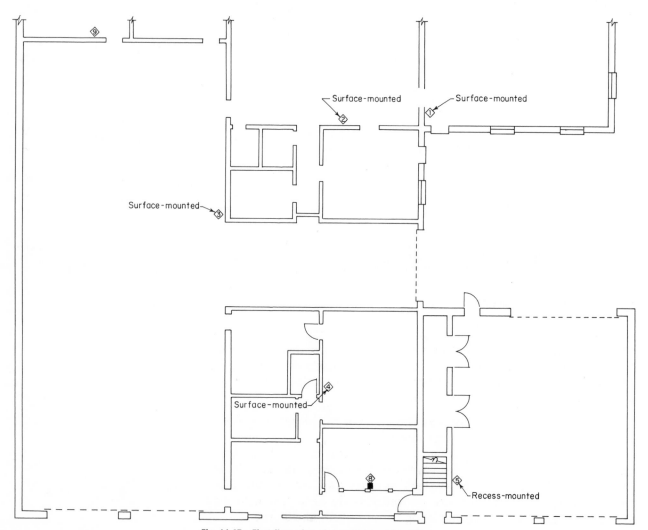

Fig. 11-27 First-floor plan of a rescue-squad building.

was decided to install a new intercom system throughout the building. The system was to be designed so that the dispatcher could talk or listen to the stations within the building either individually or simultaneously and so that any one station could call and talk with the dispatcher or any other individual station within the building.

Figures 11-27 and 11-28 show the floor plans of the building under consideration. All stations are indicated by a diamond-shaped symbol with the station number in each.

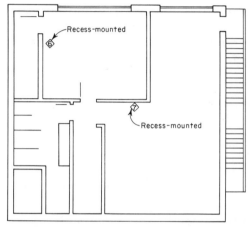

Fig. 11-28 Second-floor plan of a rescue-squad building.

The intercom-riser diagram in Fig. 11-29 gives a schematic diagram of all the stations as well as the related wiring, conduit sizes, etc. The manufacturer and catalog number of each device are also indicated on this riser diagram along with other notes, in order to facilitate the installation of the system.

A detailed description of the system should also appear in the written specifications in case the contractor wishes to substitute another brand of equipment.

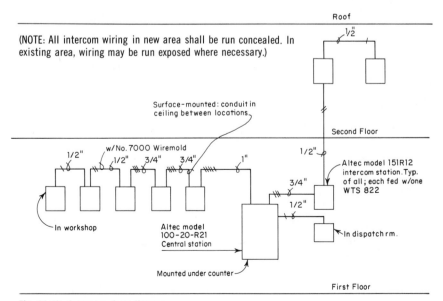

Fig. 11-29 Intercom-riser diagram.

TELEPHONE SYSTEMS

A partial floor plan of a branch bank is shown in Fig. 11-31 below. The wall-mounted telephone outlets are designated by a shaded triangle; three are shown in the illustration. Four floor-mounted telephone outlets are also shown in the teller area and are designated by a shaded triangle and enclosed in a square.

Although it is not shown on the floor plan, a telephone cabinet is located in the mechanical room of this building. The written specifications describe the type of outlet box to be used for each of the two outlet types and the telephone-riser diagram (see Fig. 11-30) shows that a ³/₄-in. empty conduit with a pull wire in each runs from each telephone outlet and terminates in the telephone cabinet.

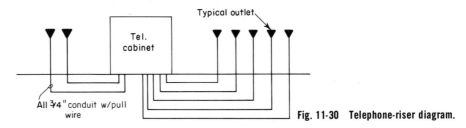

Fig. 11-30 Telephone-riser diagram.

The local telephone company is responsible for installing the telephone cabinet as well as pulling the required cable to each telephone outlet within the building.

BANK ALARM SYSTEMS

Figure 11-31 also shows various outlets for the bank security system such as camera junction boxes, smoke detectors, sound

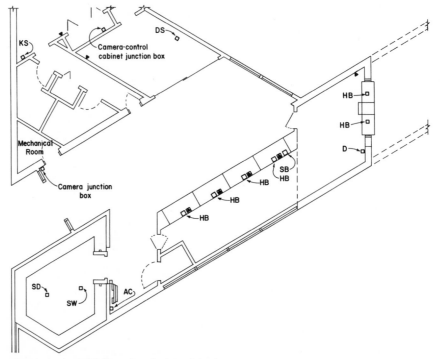

Fig. 11-31 A partial floor plan of a branch bank.

receivers, alarm buttons, etc. They are shown on the floor plan only to indicate the approximate location of each. Details of wiring are not shown in this drawing, but the riser diagram in Fig. 11-32 indicates clearly how each outlet is to be installed.

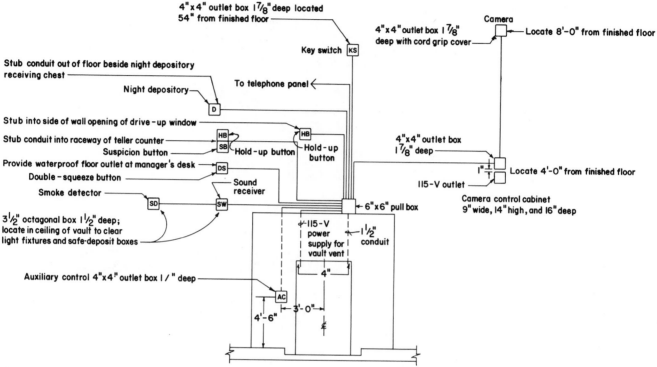

Fig. 11-32 Alarm-system riser diagram.

Chapter 12

Energy-Conservation Methods

Conservation of our natural resources through wise and efficient use of energy has now become a matter of great importance to everyone. This does not mean that the use of these resources must be stopped—returning us to the life-style of our ancestors—but it does mean that there is no place for wasteful use of energy. Everyone must strive to use available energy resources more wisely and with greater efficiency.

This new era has clearly defined the electrical engineer's role in the immediate future—to design electrical systems which will permit the use of electrical energy wisely; to see to it that these systems are installed as specified; and to instruct the system users on its proper and most efficient operation.

The following paragraphs describe various methods—employed by consulting engineers, contractors, and users—which have helped to conserve electrical energy. The use of any or all of these techniques is highly recommended whenever it is practical. Judgment must be used in each individual application to determine the feasibility of each technique.

INCANDESCENT LAMPS

Incandescent lamps should be installed only in areas where their use will be of short duration. Such areas may include clothes closets and small storage closets, janitors' closets, infrequently used storage areas, or places where the use of a more efficient light source would be impractical.

Higher-wattage general-service lamps are more efficient than lamps of a lower wattage. One 200-W lamp, for example, produces 4,010 lm as compared to 3,500 lm from two 100-W lamps. Therefore, the designer should strive to specify lighting fixtures which will handle the higher-wattage lamps. For example, the lighting fixture shown in Fig. 12-1 contains two 75-W lamps. The same style of fixture is manufactured to handle one 150-W lamp with about 15 percent increase in lamp lumen output—a much better choice when trying to conserve energy.

General-service incandescent lamps are more efficient than extended-service lamps, but at the cost of a shorter lamp life. Where the replacement of lamps causes no large problem, it is recommended that all incandescent lamps specified be of the general-service type; hard-to-reach fixtures, however, should be lamped with extended-service lamps.

If wall-wash fixtures must be used for accent lighting, try to select fluorescent or HID lighting fixtures. The color quality of certain HID and fluorescent lamps closely matches that of incandescent lamps and with a much higher efficiency (more lumens per watt).

Existing wall-wash fixtures containing incandescent lamps can be controlled by a timer switch or flasher control to turn the lights off and on at desired intervals. We have found that an interval of five seconds is about right for most applications—that is, on for five seconds, then off for five seconds. This will cut the energy consumed by 50 percent. Any duration shorter than 5 seconds becomes annoying to most, and the effect is lost for intervals longer than five seconds.

This interval delay of lighting may possibly be useful for applications other than wall-wash fixtures. In fact, for sign lighting, it would be more beneficial than a constant light since a flashing light (at short intervals) creates both light and motion which attract attention better than a constant light.

When the user leaves a room which will not be occupied again for several hours or until the next day, the lamps should, of course, be turned off unless safety reasons dictate otherwise. However, energy will not be saved by turning the lights off if the user will be returning to the room in a few minutes.

Any lighting fixture—incandescent included—should be cleaned frequently for maximum efficiency. Higher fixture efficiency will mean fewer fixtures and lamps and less wattage and wasted energy.

In rooms utilizing several incandescent fixtures for illumination, switch on the fixtures by rows because it is usually unnecessary to turn on every light in the room when natural daylight is available.

Since daylight intensity varies, switching off fixtures may not

Fig. 12-1 Illustration of a two-lamp fixture.

always be the way to conserve energy. Dimming the lights, however, may be another possibility. Incandescent lamps become very versatile tools when they are equipped with a dimming-control device; moreover, energy is saved by reducing the lamp output. Now practically all lamps, including fluorescent and HID, may be dimmed, with the possibility of conserving energy.

Since conductor and circuit-breaker sizes need not be modified by use of dimmers, existing lighting circuits are readily adaptable to dimming systems.

In addition to helping to conserve energy and lowering operating costs, dimming systems are ideal for providing security lighting and allowing selective variation in illumination levels for other multifunction areas where the lighting needs cannot always be predicted.

More emphasis should be placed on the visual task in an area than on general illumination. For example, if it is desired to have 70 fc of illumination on a desk in a small office, one fluorescent fixture mounted directly above the desk will give approximately 70 fc on the desk top while the rest of the room may average only 30 fc. However, this should not be too annoying, and almost 200 W of energy will be saved in the process.

Fluorescent Lighting

Again more concentration should be placed on the visual task than on general illumination. The floor plan of a small branch bank is shown in Fig. 12-2. Notice that sufficient, even illumina-

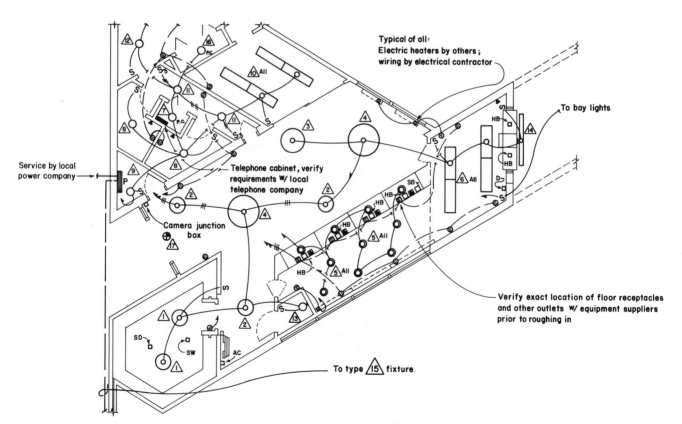

Fig. 12-2 Floor plan of a small branch bank.

tion is provided in the teller areas, but in the general lobby area uneven lighting was provided and the footcandle level varied from 20 to 70 fc. However, this gave a "warm" atmosphere within the lobby, and less energy was used than would have been if the entire lobby had been lighted to an average of 50 fc.

Fluorescent fixtures should be substituted wherever possible for incandescent lighting. The photograph in Fig. 12-3 shows an office building corridor where two-tube fluorescent fixtures had just been installed to replace recessed incandescent fixtures using single 100-W lamps. The two 20-W fluorescent lamps in the new fixtures give more lumens at half the energy of the 100-W incandescent fixtures.

Wherever practical, four-tube fluorescent fixtures should be controlled so that each ballast is switched separately. Thus, either the two inside tubes or the two outside tubes may be controlled independently. Besides offering more versatility in lighting control, half of the lamps may be out when full brilliance is not required—thus saving energy. Figure 12-4 illus-

Fig. 12-3 An office building corridor in which two-tube fluorescent fixtures have just been installed to replace recessed incandescent fixtures using a 100-W lamp in each.

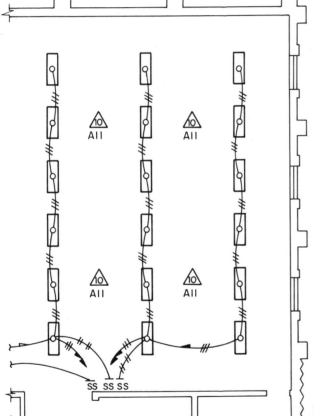

Fig. 12-4 Method used to show individual ballast switching on the working drawings.

trates how this switching arrangement is shown on an electrical working drawing.

"HEAT OF LIGHT"

Utilization of air-handling lighting fixtures is increasingly applied in a growing number of buildings and is now the preferred lighting design in many commercial structures.

Sometimes referred to as "heat of light," such fixtures utilize the heat of light as a by-product in the air-conditioning design. While saving energy in heating a building, this type of lighting fixture also has other advantages, since the removal of heat within the lighting fixture safeguards against excessive heat buildup, which can otherwise act to reduce light output and accelerate the degradation of electrical components and fixture finish materials.

Because of the cooler lamp chambers, this heat-of-light system permits higher lighting levels (up to 24 percent more light) with little or no increase in conditioned-air requirements. The result is fewer fixtures required for the visual task as well as saving in electrical energy.

Another benefit of heat of light is that it reduces the overall load on the air-conditioning equipment—another saving in energy use.

DAYLIGHTING

In the early part of this century, a distinguished French engineer by the name of Pierre Arthuys began experiments which resulted in the apparatus known as the Arthel heliostat.

A description of this sun-tapping device was published in *The Passing Show*, London (May 15, 1937). It was reportedly used to illuminate a large café in Brussels, the office building of the *Intransigeant* in Paris, the office building of the HAVOS news agency in Marseilles, the central post office at Amsterdam, and several other large offices and apartment buildings in France, Belgium, the Netherlands, North Africa, and London. All were apparently installed between 1933 and 1937.

Following is a description of heliostat use:

A huge multistory office building located in the business section of a familiar city stands lifeless in the cool morning hours before sunrise. The outline of a motionless apparatus, resembling a large mirror, can be seen on the roof as it stands out against the brighter moonlit sky. Suddenly the warmth of the first rays of morning sun strikes an ultra-sensitive thermostat located near the apparatus. This energizes an electric motor which swings the westward-facing mirror quickly over to the east. From that point the mirror is pivoted by a second motor, controlled by three other sun-operated thermostats, until the solar image is reflected in the center of the steel mirror. Then, almost instantaneously, every room in the huge building is showered with bright, diffused sunlight.

Above the mobile mirror is a large fixed mirror facing downward at an angle over a square shaft in the center of the building. The rays from the mobile mirror reflect on the fixed mirror, which deflects them in a powerful 32,000-candlepower beam straight down the shaft to the basement. Narrow shafts of light are collected from this main beam and reflected by mirrors from room to room through small apertures at ceiling level on each floor. Regardless of how far the rays must travel, there is no loss of light intensity.

If the sun is obscured by a passing cloud, the thermostat stops the motor, at the same time switching on the auxiliary electric lighting in the rooms below. When the sun comes out again, the starter-motor again goes into action, swinging the mirror forward until it catches up with the new position of the sun, which in turn switches off the electric lights in the rooms below.

General illumination within the rooms is provided by the rays of the sun passing through diffused glass panels in the ceiling while direct beams are focused at will on desks and similar seeing tasks.

This sun-tapping apparatus comes complete with several ingenious accessories. An automatic mirror-wiper prevents the upturned mirror from getting grimy. A motor-operated arm automatically sponges and polishes the mirror once a day as it swings from west to east to catch the morning sun. A pilot-thermostat controls the mirror motor, causing it to move at the same speed as the sun.

Full-time use of electrical energy offered a less expensive method of lighting buildings initially, however, and so the heliostat lost ground and was eventually discarded. It was determined, however, that this heliostat offered a saving in electrical energy of between 35 and 80 percent, depending upon the latitude in which it was used, and it may be wise for engineers to take another look at this apparatus which tapped our first light source—the sun.

While the control of daylighting is seldom within the engineer's realm of responsibility, there are certain control elements which he or she can provide.

1. Provide brightness control to prevent excessive brightness ratios.
2. Make such suggestions to the person designing the daylighting system (usually the architect) as will benefit the lighting. Such suggestions may include:
    a. Avoidance of direct sunlight
    b. Use of light reflected from the ground outside rather than the sky
    c. Use of high windows with clear glazing for deeper room illumination
    d. The use of lighting-control glass of various types for an even distribution of light

The amount of daylighting is never predictable because of the various weather conditions. For this reason, care should be taken to properly control (switch and dim) the supplemental electrical lighting fixtures which will function under a varied set of conditions.

HIGH-INTENSITY DISCHARGE (HID) LIGHTING

The increasing use of high-intensity discharge lamps for other interior lighting than industrial applications is being reported daily. Of course, there is also increasing use of HID lamps in exterior and industrial applications since the lamps have always been particularly well suited for the two latter applications.

Many developments have contributed to this increased interest in HID lamps for nonindustrial interior lighting application:

1. Great progress has been made in improving the color quality of mercury lamps to the extent that such lamps are now superior to cool-white fluorescent lamps (in some cases). The cool-white fluorescent lamp has been the number-one light source for use in commercial buildings for a number of years.

2. HID lamp life and efficiency have continually increased. For example, a new 1,000-W high-pressure sodium lamp has an initial efficiency of 130 lm per W with an average rated life of 8,000 hr on a 10-hr-per-start basis.

3. A greater range of available lamp sizes has also contributed to the increasing popularity of the HID lamp.

The use of HID lamps can be one of the greatest aids in reducing the consumption of electrical energy for lighting; some methods are as follows.

A relatively new metal halide lamp can now be operated in most indoor mercury lamp sockets and would be a good choice for many applications where the replacing of existing mercury lamps is necessary. A changeover to the higher-efficiency metal halide lamp, according to the manufacturer, will permit owners to get at least 50 percent more light initially than would be possible with mercury lamps of the same wattage. This means that the same amount of original light can be had with fewer lighting fixtures which in turn would reduce the consumption of electrical energy.

At this writing, a new 250-W high-pressure sodium lamp has also been developed for use in lamp sockets designed for 400-W mercury vapor lamps. It is claimed that this lamp will produce more initial lumens than the mercury lamp and at an obviously lower wattage.

Utility companies should use this lamp to replace old existing mercury lamps as the need arises, since this use alone would add up to a tremendous saving in electrical energy.

Those areas now using 500-W incandescent lamps as the chief light source can be dramatically improved by switching to mercury lighting. For example, if fifty 500-W incandescent lamps are needed to light a large warehouse, the incandescent fixtures could be replaced with 250-W mercury fixtures. This new system would consume only half as much energy as the incandescent lamps and still give about one-third more light output; in addition, the new mercury lamps would last about three times as long as the incandescent lamps.

Existing mercury systems can be improved by replacing inefficient mercury lamps with a relatively new type known as the Multi-Vapor. This lamp gives from 50 to 300 percent more light output than regular mercury lamps of the same wattage. Relamping existing areas with this more efficient lamp type means fewer lamps and fixtures. And a more efficient system consumes fewer watts, too.

DIRT ACCUMULATION

Dirt accumulation can severely reduce lighting levels, as can old lamps because they produce nowhere near their initial output. The usual 70 percent maintenance factor used in lighting calculations can be increased to 90 percent if commitment is made to group relamping at minimum-cost-of-light intervals, along with frequent cleanings. This can save both money and energy on the order of 20 to 40 percent.

The cases are endless, but it is now obvious that the electrical engineer or lighting designer is faced with the big task of providing sufficient light with less wasted energy. Yet we must learn to grow and progress while meeting the terms of the energy situation.

Appendix A

Electrical Specifications

The electrical specifications presented in this appendix have been successfully used for numerous commercial projects where the electrical contract has been under $100,000. Few, if any, problems have been encountered with the proper interpretation.

All projects presented in this book used this specification as a basis for the final bid specifications. Modifications are of course necessary, but this specification should be adequate for any conventional commercial electrical project, provided such items as lighting-fixture schedules, panelboard schedules, etc., are shown on the drawings.

HOW TO USE THE SPECIFICATION

1. The designer or specification writer must first become thoroughly familiar with the project: wiring methods, electrical characteristics of the electrical service, size of service, type of heating and cooling system, special equipment, emergency electrical equipment, etc.

2. The entire "master specification" is then laid out in front of the writer for review. A copy of all working drawings is also kept close at hand for reference as the master specification is being read.

3. The specification writer then begins to read the master specification, deleting any words or paragraphs not related to the given project or adding items which are necessary for the project.

4. After the master specification has been modified for the project in question, the rough drafts are usually given to a typist for retyping. This revised version is then reproduced to obtain the number of copies required for the construction documents.

EXAMPLE

The project described throughout this book as "The First National Bank—Stanley Branch" will serve to demonstrate the procedure used to write a specification. Prior to compiling the master specification as described later in this appendix, a specification for a branch bank would take at least two days to write. Using the master specification, two hours is sufficient time to get the specification ready for retyping. The following procedure is used by the designer to write the specification for the branch bank.

1. Since the designer had made all calculations, laid out all work for the draftsmen, and checked all requirements with the local utility companies, she or he was familar with the entire project prior to writing the specifications.

2. A folder containing the master specification was laid out on the work table along with a set of electrical drawings. Pads and work pencils completed the "tools" for the procedure.

3. Section 16A, General Provisions (below), was read first, and it was determined that the entire section would remain unchanged for this project.

4. Section 16B, Basic Materials and Workmanship (page 159), was the next section to be read. The designer found that it was necessary to make some changes in this section, and they were made as follows. A paragraph (number 16) was added to this section to allow for the use of armored cable; it read as follows:

> Armored cable type AC (BX) may be used in place of EMT or rigid conduit in all wood-framed partitions and ceilings where permitted by the latest edition of the National Electrical Code.

5. Section 16D was the next section to be read. It was determined that this section could remain unchanged except for subparagraph "*a*" under paragraph 7, "Distribution Panelboard—Fusible Switch (MDP)." A note was attached to this sheet to notify the typist to delete subparagraph "*a*" and change all letters after this. Therefore subparagraph "*b*" was changed to subparagraph "*a*"; subparagraph "*c*" became subparagraph "*b*," etc.

6. Section 16E, Lighting Fixtures and Lamps, remained unchanged.

7. Section 16F, Special Systems, also remained unchanged.

8. Since there were no underfloor ducts in this system, Section 16G, Underfloor Electrical Distribution System, was deleted completely.

9. The panelboard schedules were placed on the drawings for this project, and there was no need for section 16H in the specifications. Therefore, this section was deleted also.

10. There was no emergency standby electrical service nor any fire-alarm system for this project, and sections 16I and 16J were also deleted.

This completed the specifications for the project. The few sheets which had to be modified were finished and then these sheets along with the remaining sheets were sent to the offset printing room for printing.

DIVISION 16—ELECTRICAL
Section 16A—General Provisions

1. Portions of the sections of the Documents designated by the letters A, B, and C and "DIVISION ONE—GENERAL REQUIREMENTS" apply to this Division.

2. Consult Index to be certain that set of Documents and Specifications is complete. Report omissions or discrepancies to the architect.

3. SCOPE OF THE WORK

a. The scope of the work consists of the furnishing and installing of complete electrical systems—exterior and interior—including miscellaneous systems. The Electrical Contractor shall provide all supervision, labor, materials, equipment, machinery, and any and all other items necessary to complete the systems. The Electrical Contractor shall note that all items of equipment are specified in the singular; however, the Contractor shall provide and install the number of items of equipment as indicated on the drawings and as required for complete systems.

b. It is the intention of the Specifications and Drawings to call for finished work, tested, and ready for operation.

c. Any apparatus, appliance, material, or work not shown on drawings but mentioned in the specifications, or vice versa, or any incidental accessories necessary to make the work complete and perfect in all respects and ready for operation, even if not particularly specified, shall be furnished, delivered, and installed by the Contractor without additional expense to the Owner.

d. Minor details not usually shown or specified, but necessary for proper installation and operation, shall be included in the Contractor's estimate, the same as if herein specified or shown.

e. With submission of bid, the Electrical Contractor shall give written notice to the Architect of any materials or apparatus believed inadequate or unsuitable, in violation of laws, ordinances, rules, and any necessary items or work omitted. In the absence of such written notice, it is mutually agreed that the Contractor has included the cost of all required items in his proposal, and that he will be responsible for the approved satisfactory functioning of the entire system without extra compensation.

4. ELECTRICAL DRAWINGS

a. The Electrical drawings are diagrammatic and indicate the general arrangement of fixtures, equipment, and work included in the contract. Consult the Architectural drawings and details for exact location of fixtures and equipment; where same are not definitely located, obtain this information from the Architect.

b. The Contractor shall follow drawings in laying out work and check drawings of other trades to verify spaces in which work will be installed. Maintain maximum headroom and space conditions at all points. Where headroom or space conditions appear inadequate, the Architect shall be notified before proceeding with installation.

c. If directed by the Architect, the Contractor shall, without extra charge, make reasonable modifications in the layout as needed to prevent conflict with work of other trades or for proper execution of the work.

5. CODES, PERMITS, AND FEES

a. The Contractor shall give all necessary notices, including electric and telephone utilities, obtain all permits, and pay all government taxes, fees, and other costs, including utility connections or extensions, in connection with his work; file all necessary plans, prepare all documents, and obtain all necessary approvals of all governmental departments having jurisdiction; obtain all required certificates of inspection for his work and deliver same to the Architect before request for acceptance and final payment for the work.

b. The Contractor shall include in the work, without extra cost to the Owner, any labor, materials, services, apparatus, drawings (in addition to contract drawings and documents) in order to comply with all applicable laws, ordinances, rules, and regulations, whether or not shown on drawings and/or specified.

c. Work and materials shall conform to the latest rules of the National Board of Fire Underwriters' Code, Regulations of the State Fire Marshal, and with applicable local codes and with all prevailing rules and regulations pertaining to adequate protection and/or guarding of any moving parts, or otherwise hazardous conditions. Nothing in these specifications shall be construed to permit work not conforming to the most stringent of applicable codes.

d. The National Electrical Code, the local electric code, and the electrical requirements as established by the State and Local Fire Marshal, and rules and regulations of the power company serving the project, are hereby made part of this specification. Should any changes be necessary in the drawings or specifications to make the work comply with these requirements, the Electrical Contractor shall notify the Architect.

6. SHOP DRAWINGS

a. The Electrical Contractor shall submit five (5) copies of the shop drawings to the Architect for approval within thirty (30) days after the award of the general contract. If such a schedule cannot be met, the Electrical Contractor may request in writing for an extension of time to the Architect. If the Electrical Contractor does not submit shop drawings in the prescribed time, the Architect has the right to select the equipment.

b. Shop drawings shall be submitted on all major pieces of electrical equipment, including service-entrance equipment, lighting fixtures, panelboards, switches, wiring devices and plates, and equipment for miscellaneous systems. Each item of equipment proposed shall be a standard catalog product of an established manufacturer. The shop drawing shall give complete information on the proposed equipment. Each item of the shop drawings shall be properly labeled, indicating *the intended service of the material*, the job name, and Electrical Contractor's name.

c. The shop drawings shall be neatly bound in five (5) sets and submitted to the Architect with a letter of transmittal. The letter of transmittal shall list each item submitted along with the manufacturer's name.

d. Approval rendered on shop drawings shall not be considered as a guarantee of measurements or building conditions. Where drawings are approved, said approval does not mean that drawings have been checked in detail; said approval does not in any way relieve the Contractor from his responsibility, or necessity of furnishing material or performing work as required by the contract drawings and specifications.

7. COOPERATION WITH OTHER TRADES

a. The Electrical Contractor shall give full cooperation to other trades and shall furnish (in writing, with copies to the Architect) any information necessary to permit the work of all trades to be installed satisfactorily and with least possible interference or delay.

b. Where the work of the Electrical Contractor will be installed in close proximity to work of other trades, or where there is evidence that the work of the Electrical Contractor will interfere with the work of other trades, he shall assist in working out space conditions to make a satisfactory adjustment. If so directed by the Architect, the Electrical Contractor shall prepare composite working drawings and sections at a suitable scale clearly showing other trades. If the Electrical Contractor installs his work before coordinating with work of other trades, he shall make necessary changes in his work to correct the condition without extra charge.

c. The complexity of equipment and the variation between equipment manufacturers require complete coordination of all trades. The Contractor, who offers, for consideration, substitutes of equal products of reliable manufacturers, has to be responsible for all changes that affect his installation and the installation and equipment of other trades. All systems and their associated controls must be completely installed, connected, and operating to the satisfaction of the Architect prior to final acceptance and contract payment.

8. TEMPORARY ELECTRICAL SERVICE

a. The Electrical Contractor shall be responsible for all arrangements and costs for providing, at the site, temporary electrical metering, main switches, and distribution panels as required for construction purposes. The distribution panels shall be located at a central point designated by the Architect. The General Contractor shall indicate prior to installation whether three-phase or single-phase service is required.

9. ELECTRICAL CONNECTIONS

a. The Electrical Contractor shall provide and install power wiring to all motors and electrical equipment complete and ready for operation including disconnect switches and fuses. Starters, relays, and accessories shall be furnished by others unless otherwise noted, but shall be installed by the Electrical Contractor. This Contractor shall be responsible for checking the shop drawings of the equipment manufacturer to obtain the exact location of the electrical rough-in and connections for equipment installed.

b. The Mechanical Contractor will furnish and install all temperature-control wiring and all interlock wiring unless otherwise noted.

c. It shall be the responsibility of the Electrical Contractor to check all motors for proper rotation.

10. AS-BUILT DRAWINGS

a. The Electrical Contractor shall maintain accurate records of all deviations in work as actually installed from work indicated on the drawings. On completion of the project, two (2) complete sets of marked-up prints shall be delivered to the Architect.

11. INSPECTION AND CERTIFICATES

a. On the completion of the entire installation, the approval of the Architect and Owner shall be secured, covering the installation throughout. The Contractor shall obtain and pay for Certificate of Approval from the public authorities having jurisdiction. A final inspection certificate shall be submitted to the Architect prior to final payment. Any and all costs incurred for fees shall be paid by the Contractor.

12. TESTS

a. The right is reserved to inspect and test any portion of the equipment and/or materials during the progress of its erection. The Contractor shall test all wiring and connections for continuity and grounds, before connecting any fixtures or equipment.

b.   The Contractor shall test the entire system in the presence of the Architect or his engineer when the work is finally completed, to ensure that all portions are free from short circuits or grounds. All equipment necessary to conduct these tests shall be furnished at the Contractor's expense.

13.   EQUIVALENTS

a.   When material or equipment is mentioned by name, it shall form the basis of the Contract. When approved by the Architect in writing, other material and equipment may be used in place of those specified, but written application for such substitutions shall be made to the Architect as described in the Bidding Documents. The difference in cost of substitute material or equipment shall be given when making such request. Approval of substitute is, of course, contingent on same meeting specified requirements and being of such design and dimensions as to comply with space requirements.

14.   GUARANTEE

a.   The Electrical Contractor shall guarantee, by his acceptance of the Contract, that all work installed will be free from defects in workmanship and materials. If during the period of one year, or as otherwise specified, from date of Certificate of Completion and acceptance of work, any such defects in workmanship, materials, or performance appear, the Contractor shall, without cost to the Owner, remedy such defects within a reasonable time to be specified in notice from the Architect. In default, the Owner may have such work done and charge cost to Contractor.

Section 16B — Basic Materials and Workmanship

1.   Portions of the sections of the Documents designated by the letters A, B, and C and "DIVISION ONE — GENERAL REQUIREMENTS" apply to this Division.

2.   Consult Index to be certain that set of Documents and Specifications is complete. Report omissions or discrepancies to the Architect.

3.   CONDUIT MATERIAL AND WORKMANSHIP

a. *General*   The Electrical Contractor shall install a complete raceway system as shown on the drawings and stated in other sections of the specifications. All material used in the raceway system shall be new and the proper material for the job. Conduit, couplings, and connectors shall be a product of a reputable manufacturer equal to conduit as manufactured by Triangle Conduit and Cable or National Electric.

b. *Conduit installation*

(1)   Conduit shall be of ample size to permit the ready insertion and withdrawal of conductors without abrasion. All joints shall be cut square, reamed smooth, and drawn up tight.

(2)   Concealed conduits shall be on run in as direct a manner and with as long a bend as possible. Exposed con-

duit shall be run parallel to or at right angles with the lines of the building. All bends shall be made with standard ells, conduit bent to a radius not less than shown in NEC, or screw-jointed conduit fittings. All bends shall be free of dents or flattening. Not more than the equivalent of four quarter-bends shall be used in any run between terminals and cabinets, or between outlets and junction or pull boxes.

(3) Conduits shall be continuous from outlet to outlet and from outlet to cabinets, or junction or pull boxes, and shall enter and be secured at all boxes in such a manner that each system shall be electrically continuous throughout.

(4) A No. 14 galvanized iron or steel fish wire shall be left in all conduits in which the permanent wiring is not installed.

(5) Where conduits cross building joints, furnish and install O.Z. Electric Manufacturing Company expansion fittings for contraction, expansion, and settlement.

(6) Open ends shall be capped with approved manufactured conduit seals as soon as installed and kept capped until ready to install conductors.

(7) Conduit shall be securely fastened to all sheet metal outlets, junction, and pull boxes with galvanized lock-nuts and bushings, care being observed to see that the full-number threads project through to permit the bushings to be drawn tight against the end of conduit, after which the lock-nut shall be made up sufficiently tight to draw the bushings into firm electrical contact with the box.

(8) For all flush-mounted panels there shall be provided and installed 1¼-in. empty conduit up through the wall and turned out above the ceiling and one 1¼-in. empty conduit down to space below floor except where slab in on grade.

c. *Conduit hangers and supports*

(1) Conduit throughout the project shall be securely and rigidly supported to the building structure in a neat and workmanlike manner, and wherever possible, parallel runs of horizontal conduit shall be grouped together on adjustable trapeze hangers. Support spacing shall not be more than eight (8) feet.

(2) Exposed conduit shall be supported by one-hole malleable iron straps, two-hole straps, suitable beam clamps, or split-ring conduit hanger with support rod.

(3) Single conduit 1¼ inches and larger run concealed horizontally shall be supported by suitable beam clamps or split-ring conduit hangers with support rod. Multiple runs of conduit shall be grouped together on trapeze hangers where possible. Vertical runs shall be supported by steel riser clamps.

(4) Conduit one inch and smaller run concealed above a ceiling may be supported directly to the building structure with strap hangers or No. 14 gauge galvanized wire provided the support spacing does not exceed four (4) feet.

4. OUTLET BOXES

a. *General*

(1) Before locating the outlet boxes, check all of the architectural drawings for type of construction and to make sure

that there is no conflict with other equipment. The outlet boxes shall be symmetrically located according to room layout and shall not interfere with other work or equipment. Also note any detail of the outlets shown on the drawings.

(2) Outlet boxes shall be made of galvanized sheet steel unless otherwise noted or required by the NEC and shall be of the proper code size for the required number of conductors. Outlet boxes shall be a minimum of four (4) inches square unless specifically noted on the drawings with the exception that a box containing only two current-carrying conductors may be smaller. The outlet boxes shall be complete with the approved type of connectors and required accessories.

(3) The outlet boxes shall be complete with raised device covers as required to accept device installed. All outlet boxes must be securely fastened in position with the exposed edge of the raised device cover set flush with the finished surface. Approved factory-made knockout seals shall be installed where knockouts are not intact. Galvanized outlet boxes shall be manufactured by Raco, Steel City, Appleton, or approved equal.

(4) Outlet boxes for exposed work shall be handy boxes with handy box covers unless otherwise noted.

(5) Outlet boxes located on the exterior in damp or wet locations or as otherwise noted shall be threaded cast-aluminum device boxes such as Crouse Hinds type FS or FD.

b. Receptacle outlet boxes: Wall receptacles shall be mounted approximately 18 inches above the finished floor (AFF) unless otherwise noted. When the receptacle is mounted in a masonry wall, the bottom of the outlet box shall be in line with the bottom of a masonry unit. Receptacles for electric water coolers shall be installed behind the coolers in accordance with manufacturer's recommendations. All receptacle outlet boxes shall be equipped with grounding lead which shall be connected to grounding terminal of the device.

c. Switch outlet boxes: Wall switches shall be mounted approximately 54 inches above the finished floor (AFF) unless otherwise noted. When the switch is mounted in a masonry wall, the bottom of the outlet box shall be in line with the bottom of a masonry unit. Where more than two switches are located, the switches shall be mounted in a gang outlet box with gang cover. Dimmer switches shall be individually mounted unless otherwise noted. Switches with pilot lights, switches with overload motor protection, and other special switches that will not conveniently fit under gang wall plates may be individually mounted.

d. Lighting-fixture outlet boxes: The lighting-fixture outlet boxes shall be furnished with the necessary accessories to install the fixture. The supports must be such as not to depend on the outlet box supporting the fixture. The supports for the lighting fixture shall be independent of the ceiling system. All ceiling outlet boxes shall be equipped with raised circular cover plates with edge set flush with surface of finished ceiling.

5. PULL BOXES

   a. Pull boxes shall be installed at all necessary points, whether indicated on the drawings or not, to prevent injury to the insulation or other damage that might result from pulling resistance for other reasons necessary to proper installation. Pull box locations shall be approved by the Architect prior to installation. Minimum dimensions shall be not less than NEC requirements and shall be increased if necessary for practical reasons or where required to fit a job condition.

   b. All pull boxes shall be constructed of galvanized sheet steel, code gauge, except that no less than 12 gauge shall be used for any box.

   c. Where boxes are used in connection with exposed conduit, plain covers attached to the box with a suitable number of countersunk flathead machine screws may be used.

   d. Where so indicated, certain pull boxes shall be provided with barriers. These pull boxes shall have a single cover plate, and the barriers shall be of the same gauge as the pull box.

   e. Each circuit in pull box shall be marked with a tag guide denoting panels to which they connect.

   f. Exposed pull boxes will not be permitted in the public spaces.

6. WIREWAYS OR WIRE TROUGHS

   a. Wireways shall be used where indicated on the drawings and for mounting groups of switches and/or starters. Wireways shall be the standard manufactured product of a company regularly producing wireway and shall not be a local shop-assembled unit. Wireway shall be of the hinged-cover type, Underwriters' listed, and of sizes indicated or as required by NEC. Finish shall be medium light-gray enamel over rust inhibitor. Wireways shall be of raintight construction where required. Wireways shall be General Electric type HS or approved equal.

7. CONDUCTOR MATERIAL AND WORKMANSHIP

   a. *General*

      (1) The Electrical Contractor shall provide and install a complete wiring system as shown on the drawing or specifications herein. All conductors used in the wiring system shall be soft-drawn copper wire having a conductivity of not less than 98 percent of that of pure copper, with 600-volt rating, unless otherwise noted. Wire shall be as manufactured by General Cable, Triangle, or approved equal.

      (2) The wire shall be delivered to the site in its original unbroken packages, plainly marked or tagged as follows: (a) Underwriters' labels; (b) size, kind, and insulation of the wire; (c) name of manufacturing company and the trade name of the wire.

   b. *Conductor workmanship*

      (1) Install conductors in all raceways as required, unless otherwise noted, in a neat and workmanlike manner. Telephone conduits and empty conduits, as noted, shall have a No. 14 gauge galvanized pull wire left in place for future use.

      (2) Conductors shall be color-coded in accordance with the National Electrical Code. Mains, feeders, and subfeeders shall be tagged in all pull, junction, and outlet boxes and in

the gutter of panels with approved code type wire markers.
(3) No lubricant other than powdered soapstone or approved pulling compound may be used to pull conductors.
(4) At least eight (8) inches of slack wire shall be left in every outlet box whether it be in use or left for future use.
(5) All conductors and connections shall test free of grounds, shorts, and opens before turning the job over to the Owner.

8. LUGS, TAPS, AND SPLICES

a. Joints on branch circuits shall occur only where such circuits divide and shall consist of one through circuit to which shall be spliced the branch from the circuit. In no case shall joints in branch circuits be left for the fixture hanger to make. No splices shall be made in conductor except at outlet boxes, junction boxes, or splice boxes.

(1) All joints or splices for No. 10 AWG or smaller wire shall be made with UL-approved wire nuts or compression-type connectors.

(2) All joints or splices for No. 8 AWG or larger wire shall be made with a mechanical compression connector. After the conductors have been made mechanically and electrically secure, the entire joint or splice shall be covered with Scotch No. 33 tape or approved equal to make the insulation of the joint or splice equal to the insulation of the conductors. The connector shall be UL approved.

9. ACCESS DOORS

a. The Electrical Contractor shall furnish to the lather the access doors as shown on the drawings or required for access to junction boxes, etc. The doors shall be 12-inch-square, unless otherwise noted, hinged metal doors with metal frames.

b. Door and frame shall be not lighter than 16-gauge sheet steel. The access door shall be of the flush type with screwdriver latching device. The frame shall be constructed so that it can be secured to building material as required. The access doors shall be Milcor or equal. Access door and location shall meet the approval of the Architect.

10. FUSES

a. Fuses manufactured by Buss or Shawmut shall be furnished and installed as required. Motor protection fuses shall be dual-element.

11. CUTTING AND PATCHING

a. On new work the Electrical Contractor shall furnish sketches to the General Contractor showing the locations and sizes of all openings and chases, and furnish and locate all sleeves and inserts required for the installation of the electrical work before the walls, floors, and roof are built. The Electrical Contractor shall be responsible for the cost of cutting and patching where any electrical items were not installed or where incorrectly sized or located. The Contractor shall do all drilling required for the installation of his hangers.

b. On alterations and additions to existing projects, the Electrical Contractor shall be responsible for the cost of all cutting and patching, unless otherwise noted.

c. No structural members shall be cut without the approval of the Architect, and all such cutting shall be done in a manner directed by him. All patching shall be performed in a neat and workmanlike manner acceptable to the Architect.

12. EXCAVATION AND BACKFILLING

a. The Electrical Contractor shall be responsible for excavation, backfill, tamping, shoring, bracing, pumping, street cuts, repairing of finished surface, and all protection for safety of persons and property as required for installing a complete electrical system. All excavation and backfill shall conform to the Architectural Section of the specifications.

b. Excavation shall be made in a manner to provide a uniform bearing for conduit. Where rock is encountered, excavate three (3) inches below conduit grade and fill with gravel to grade.

c. After required test and inspections, backfill the ditch and tamp. The first foot above the conduit shall be hand-backfilled with rockfree clean earth. The backfill in the ditches on the exterior and interior of the building shall be tamped to 90 percent. The Electrical Contractor shall be responsible for any ditches that go down.

13. EQUIPMENT AND INSTALLATION WORKMANSHIP

a. All equipment and material shall be new and shall bear the manufacturer's name and trade name. The equipment and material shall be essentially the standard product of a manufacturer regularly engaged in the production of the required type of equipment and shall be the manufacturer's latest approved design.

b. The Electrical Contractor shall receive and properly store the equipment and material pertaining to the electrical work. The equipment shall be tightly covered and protected against dirt, water, chemical or mechanical injury, and theft. The manufacturer's directions shall be followed completely in the delivery, storage, protection, and installation of all equipment and materials.

c. The Electrical Contractor shall provide and install all items necessary for the complete installation of the equipment as recommended or as required by the manufacturer of the equipment or required by local code without additional cost to the Owner, regardless of whether the items are shown on the plans or covered in the specifications.

d. It shall be the responsibility of the Electrical Contractor to clean the electrical equipment, make necessary adjustments, and place the equipment into operation before turning equipment over to the Owner. Any paint that was scratched during construction shall be "touched-up" with factory-color paint to the satisfaction of the Architect. Any items that were damaged during construction shall be replaced.

14. CONCRETE PADS, SUPPORTS, AND ENCASEMENT

a. The Electrical Contractor shall be responsible for all concrete pads, supports, piers, bases, foundations, and encasement required for the electrical equipment and conduit. The concrete pads for the electrical equipment shall be six (6) inches larger all around than the base of the equipment and a minimum of four (4) inches thick unless specifically indicated otherwise.

15. WATERPROOFING

a. The Electrical Contractor shall provide all flashing,

caulking, and sleeves required where his items pass through the outside walls or roof. The waterproofing of the openings shall be made absolutely watertight. The method of installation shall conform to the requirements of Division 7, "Moisture Control," and/or meet the approval of the Architect.

Section 16C—Service-Entrance System

1. Portions of the sections of the Documents designated by the letters A, B, and C and "DIVISION ONE—GENERAL REQUIREMENTS" apply to this Division.

2. Consult Index to be certain that set of Documents and Specifications is complete. Report omissions or discrepancies to the Architect.

3. SERVICE ENTRANCE

a. General: The Electrical Contractor shall provide and install a complete service-entrance system as shown on the drawings or as required for a complete system. All material and workmanship shall conform with Section 16B of the specifications, National Electrical Code, and the local electric code. The electric service entrance shall conform to the requirements and regulations of the electric utility serving the project.

b. Electric utility charge: The Electrical Contractor shall make all arrangements with the electric utility and pay all charges made by the electric utility for permanent electric service to the project. In the event that the charges of the electric utility are not available at the time the project is bid, the Electrical Contractor shall qualify his bid to notify the Owner that such charges are not included.

c. Metering: The Electrical Contractor shall provide and install raceway, install current transformer cabinet, and/or meter trim, for metering facilities as required by the electric utility serving the project. The electric utility will provide the meter installation including meter, current transformers, and connections.

d. Grounding: The Electrical Contractor shall properly ground the electrical system as required by the National Electrical Code. The ground wire for the service entrance shall be run in conduit and made to the main water service and connected ahead of any valve or cutoff.

e. Conduit: The conduit used for service entrance shall be galvanized rigid steel conduit unless otherwise noted on drawings.

f. Conductors: Conductors for the service entrance shall be copper type RHW or THW rated at 75°C unless otherwise noted. The conductors indicated on the drawings are based on copper. Conductors with a size of No. 1/0 and larger may be aluminum, provided the size of the conductor is increased to have the same current-carrying capacity as the copper conductors (or more). Also, the conduit size shall be increased accordingly.

Section 16D—Electrical Distribution System

1. Portions of the sections of the Documents designated by the letters A, B, and C and "DIVISION ONE—GENERAL REQUIREMENTS" apply to this Division.

2. Consult Index to be certain that set of Documents and Specifications is complete. Report omissions or discrepancies to the architect.

3. FEEDERS AND BRANCH CIRCUITS

a. General: The Electrical Contractor shall provide and install a complete electrical distribution system as shown on the drawings or as required for a complete system. All materials and workmanship shall conform with Section 16B of the Specifications, National Electrical Code, and the local electric code.

b. Conduit materials:

(1) Rigid conduit (heavy wall): Rigid conduit shall be galvanized rigid steel conduit with a minimum size of ³/₄ inch unless otherwise noted. Rigid steel conduit shall be installed for the following services and locations: service entrance, underground in contact with earth, in concrete slab, panel feeders, exterior of building walls, motor feeders over 10 hp, electrical equipment feeders over 10 kW, "wet" locations, and as required by the National Electrical Code and local codes.

(2) Electrical metallic tubing (EMT): Electrical metallic tubing shall be galvanized steel with a minimum size of ³/₄ inch. Electrical metallic tubing shall be used in all locations not otherwise specified for rigid or flexible conduit and where not in violation of the National Electrical Code.

(3) Flexible metal conduit: Flexible metal conduit shall be galvanized steel. Flexible metal conduit located in wet locations shall be the Liquid-Tight type. Flexible metal conduit may be used in place of EMT where completely accessible, such as above removable acoustical tile ceilings and for exposed work in unfinished spaces.

(4) A short piece of flexible metal conduit shall be used for the connection to all motors and vibrating equipment, connection between recessed light fixtures and junction box, and as otherwise noted, provided the use meets the requirements of the National Code and local codes. The flexible metal conduit shall be the type approved for continuous grounding.

c. Conductor material:

(1) The conductor material shall be as follows, unless otherwise noted:

(a) *Feeders:* Shall be Type RHW or THW rated at 75°C.

(b) *Branch Circuits:* Shall be Type THW rated at 75°C, except that branch circuits with conductor sizes of No. 10 and smaller in dry locations may be Type TW rated at 60°C.

(c) *Special Locations:* Conductors in special locations such as range hoods, lighting fixtures, etc., shall be as required by the National Electrical Code, local code, or as otherwise noted.

(2) No conductor shall be smaller than No. 12 wire, except for the control wiring and as stated in other sections of the Specifications or on the drawings. Wiring to switches shall not be considered as control wiring.

(3) Conductors indicated on the drawings are based on copper. Panel, motor, and electrical equipment feeders with

a size of No. 1/0 and larger may be aluminum, provided the size of the conductor is increased to have the same current-carrying capacity as the copper conductors (or more). Also, the conduit sizes shall be increased accordingly.

(4) All conductors with the size of No. 8 or larger shall be stranded.

(5) All lighting and receptacle branch circuits in excess of 100 linear feet shall be increased one size to prevent excessive voltage drop.

4. SAFETY SWITCHES (FSS) (NFSS)

a. General: Furnish and install safety switches as indicated on the drawings or as required. All safety switches shall be *NEMA General Duty Type* and Underwriters' Laboratories Listed. The switches shall be Fused Safety Switches (FSS) or Nonfused Safety Switches (NFSS) as shown on the drawings or required.

b. Switches: Switches shall have a quick-make and quick-break operating handle and mechanism which shall be an integral part of the box. Padlocking provisions shall be provided for padlocking in the OFF position with at least three padlocks. Switches shall be horsepower rated for 250 volts ac or dc as required. Lugs shall be UL-listed for copper and aluminum cable.

c. Enclosures: Switches shall be furnished in NEMA 1 general-purpose enclosures with knockouts unless otherwise noted or required. Switches located on the exterior of the building or in "wet" locations shall have NEMA 3R enclosures (WP).

d. Installation: The safety switches shall be securely mounted in accordance with the NEC. The Contractor shall provide all mounting materials. Install fuses in the FSS. The fuses shall be dual-element on motor circuits.

e. Manufacturer: Square "D," General Electric, Cutler-Hammer, or Westinghouse.

5. PANELBOARDS—CIRCUIT BREAKER

a. General: Furnish and install circuit-breaker panelboards as indicated in the panelboard schedule and where shown on the drawings. The panelboard shall be dead-front safety type equipped with molded-case circuit breakers and shall be the type as listed in the panelboard schedule. Service-entrance panelboards shall include a full-capacity box bonding strap and be approved for service entrance. The acceptable manufacturers of the panelboards are Square "D," General Electric, Cutler-Hammer, and Westinghouse provided they are fully equal to the type listed on the drawings. The panelboard shall be listed by Underwriters' Laboratories and bear the UL label.

b. Circuit breakers: Provide molded-case circuit breakers of frame, trip rating, and interrupting capacity as shown on the schedule. Also, provide the number of spaces for future circuit breakers as shown in the schedule. The circuit breakers shall be quick-make, quick-break, thermal-magnetic, trip-indicating, and have common trip on all multipole breakers with internal tie mechanism.

c. Panelboard bus assembly: Bus bar connections to the branch circuit breakers shall be the "phase-sequence" type. Single-phase three-wire panelboard bussing shall be such that any two adjacent single-pole breakers are connected to opposite polarities in such a manner that two-pole breakers can be installed in any location. Three-phase four-wire bussing shall be such that any three adjacent single-pole breakers are individually connected to each of the three different phases in such a manner that two-or three-pole breakers can be installed at any location. All current-carrying parts of the bus assembly shall be plated. Mains ratings shall be as shown in the panelboard schedule on the plans. Provide solid neutral (S/N) assembly when required.

d. *Wiring terminals:* Terminals for feeder conductors to the panelboard mains and neutral shall be suitable for the type of conductor specified. Terminals for branch-circuit wiring, both breaker and neutral, shall be suitable for the type of conductor specified.

e. *Cabinets and fronts:* The panelboard bus assembly shall be enclosed in a steel cabinet. The size of the wiring gutters and gauge of steel shall be in accordance with NEMA Standards. The box shall be fabricated from galvanized steel or equivalent rust-resistant steel. Fronts shall include door and have flush, brushed stainless steel, spring-loaded door pulls. The flush lock shall not protrude beyond the front of the door. All panelboard locks shall be keyed alike. Fronts shall not be removable with door in the locked position.

f. *Directory:* On the inside of the door of each cabinet, provide a typewritten directory which will indicate the location of the equipment or outlets supplied by each circuit. The directory shall be mounted in a metal frame with a nonbreakable transparent cover. The panelboard designation shall be typed on the directory card and panel designation stenciled in 1½-inch-high letters on the inside of the door.

g. *Panelboard installation:*

(1) Before installing panelboards, check all the architectural drawings for possible conflict of space, and adjust the location of the panelboard to prevent such conflict with other items.

(2) When the panelboard is recessed into a wall serving an area with accessible ceiling space, provide and install an empty conduit system for future wiring. A 1¼-inch conduit shall be stubbed into the ceiling space above the panelboard and under the panelboard if such accessible ceiling space exists.

(3) The panelboards shall be mounted in accordance with Article 373 of the NEC. The Electrical Contractor shall furnish all material for mounting the panelboards.

6. WIRING DEVICES

a. *General:* The wiring devices specified below with Arrow Hart numbers may also be the equivalent wiring device as manufactured by Bryant Electric, Harvey Hubbell, or Pass & Seymour. All other items shall be as specified.

b. *Wall switches:* Where more than one flush wall switch is indicated in the same location, the switches shall be mounted in gangs under a common plate.

(1)	Single-pole	AH#1991
(2)	Three-way	AH#1993
(3)	Four-way	AH#1994
(4)	Switch with pilot light	AH#2999-R
(5)	Motor switch—surface	AH#6808
(6)	Motor switch—flush	AH#6808-F

c. *Receptacles:*

(1)	Duplex	AH#6739
(2)	Clock outlet	AH#5708
(3)	Weatherproof	AH#5735-F

(4) Floor receptacles: Steel City Series 600 floor box with bronze edge ring, floor plate P-60-1, bronze carpet plate, and service fitting SFH-40-RG.

(5) Floor outlet for telephone and alarm: Steel City Series 600 floor box with bronze edge ring, floor plate P-60-1, bronze carpet plate, and service fitting SFL-10.

d. *Wall plate:* Stainless steel wall plates with satin finish minimum 0.030 inch shall be provided for all outlets and switches.

7. DISTRIBUTION PANELBOARD—FUSIBLE SWITCH (MDP)

a. *General:* Furnish and install distribution and power panelboards as indicated in the panelboard schedule and where shown on the drawings. Panelboards shall be dead-front safety type, equipped with quick-make, quick-break fusible branch switches, and approved for service entrance. The acceptable manufacturers of the panelboard are Square "D," General Electric, Cutler-Hammer, and Westinghouse provided they are fully equal to the type listed on the drawings. The panelboard shall be listed by Underwriters' Laboratories and bear the UL label.

b. *Fusible switches:* All fusible branch switches shall be quick-make, quick-break, with visible blades and dual horsepower ratings. Switch handles shall physically indicate ON and OFF positions. Such handles shall also be able to accept three padlocks having heavy-duty industrial-type shackles. Covers shall be interlocked with the switch handles to prevent opening in the ON position. A means shall be provided to allow authorized personnel to release the interlock for inspection purposes when a switch is ON. A cardholder, providing circuit identification, shall be mounted on each branch switch. Switches shall be provided with Bussman Fusetron fuses or as noted on the drawings.

c. *Bussing assembly:* Panelboard bus structure and main lugs or main switch shall have current ratings as shown on the panelboard schedule. The bus structure shall accommodate plug-on or bolted branch switches and motor starters as indicated in the panelboard schedule without modification to the bus assembly. Provide solid neutral (S/N) assembly when required.

d. *Equipment rating:* Switches and panelboard bus structure shall safely and without failure withstand short circuits on the systems capable of delivering up to 50,000 amperes RMS symmetrical, unless otherwise noted.

e. *Cabinet:* Panelboard assembly shall be enclosed in a steel cabinet. The rigidity and gauge of steel is to be as specified in UL Standard for cabinets. The size of wiring gutters shall be in accordance with UL Standard. Cabinets shall be equipped with a front door and have fully concealed, self-aligning trim clamps. Fronts shall be full-finished steel with rust-inhibiting primer and baked enamel finish.

f. Wiring terminals: Terminals for feeder conductors to the panelboard mains and neutral shall be suitable for the type of conductor specified. Terminals for branch-circuit wiring, both breaker and neutral, shall be suitable for the type of conductor specified.

g. Panelboard installation: Before installing panelboards, check all of the Architectural drawings for possible conflict of space and adjust the location of the panelboard to prevent such conflict with other items. The panelboards shall be mounted in accordance with the NEC. The Electrical Contractor shall furnish all material for mounting the panelboards.

Section 16E — Lighting Fixtures and Lamps

1. Portions of the sections of the documents designated by the letters A, B, and C and "DIVISION ONE — GENERAL REQUIREMENTS" apply to this Division.

2. Consult Index to be certain that set of Documents and Specifications is complete. Report omissions or discrepancies to the Architect.

3. LIGHTING FIXTURES

a. General: The Electrical Contractor shall furnish, install, and connect all lighting fixtures to the building wiring system unless otherwise noted.

b. Fixture type: The fixture for each location is indicated by type letter. Refer to fixture schedule on the drawings for each type, manufacturer, catalog number, and type of mounting.

c. Fluorescent ballasts: All fluorescent fixtures shall have ETL-CBM high power factor, quiet-operating, Class A sound-rated, thermally protected Class P cool-rated ballast with UL approval. Ballasts shall be as manufactured by General Electric, Advance, Jefferson or approved equal. The ballasts shall be subject to a two (2) year manufacturer's guarantee. Guarantee shall be filed with the Owner.

d. Shop drawings:

(1) Shop drawings for lighting fixtures shall indicate each type together with manufacturer's name and catalog number, complete photometric data compiled by an independent testing laboratory, and type of lamp(s) to be installed. No fixtures shall be delivered to the job until approved by the Architect.

(2) If the Electrical Contractor submits shop drawings on a fixture for approval other than those specified, he shall also submit a sample fixture when requested by the Architect. The sample fixture will be returned to the Electrical Contractor. The decision of the Architect shall be final.

e. Coordination: It shall be the responsibility of the Electrical Contractor to coordinate with the ceiling contractor and the General Contractor in order that the proper type of fixture be furnished to match the ceiling suspension system being installed or building construction material.

4. LAMPS

   a. The Electrical Contractor shall furnish and install lamps in all fixtures as indicated on the drawings or as required. Fluorescent lamps shall be standard cool white, and incandescent lamps shall be inside-frosted unless otherwise noted on the drawings.

   b. Lamps shall be manufactured by General Electric, Westinghouse, or Sylvania.

Section 16F—Special Systems

1. Portions of the sections of the Documents designated by the letters A, B, and C and "DIVISION ONE—GENERAL REQUIREMENTS" apply to this Division.

2. Consult Index to be certain that set of Documents and Specifications is complete. Report omissions or discrepancies to the Architect.

3. TELEPHONE RACEWAY SYSTEMS

   a. *General:* The Electrical Contractor shall provide and install empty raceway, outlet boxes, pull boxes, and associated equipment required for a complete telephone system as indicated on the drawings and specified herein. All materials and workmanship shall conform with Section 16B of the Specifications. All wiring shall be installed by the local telephone company. The entire installation shall be in accordance with the requirements of the local telephone company.

   b. *Rigid conduit* (heavy wall): Rigid conduit shall be installed in the following locations: Service entrance, underground in contact with earth, in concrete slab and "wet" locations.

   c. *Electric metallic conduit (EMT):* Electric metallic tubing shall be used in all locations not otherwise specified to be rigid conduit.

   d. *Outlets:* Telephone wall outlets shall consist of a 4-inch two-gang outlet box, raised device cover, and a telephone device plate of the same material as the receptacle device plates. The conduit shall extend from the outlet to the designated telephone space unless otherwise noted.

   e. *Pull wire:* The Electrical Contractor shall install a No. 14 gauge galvanized pull wire in the raceway system for future use.

   f. *Mounting heights:* The wall outlets shall be mounted at approximately the following heights unless otherwise noted on the drawings or required by the telephone company: Desk Phones—18 inches AFF; Wall Phone—58 inches AFF; Telephone Booth—7 feet-6 inches AFF.

   g. *Floor outlets:* See wiring devices, Section 16D. Verify exact location with Architect.

4. EMERGENCY LIGHTING SYSTEM

   a. The Electrical Contractor shall provide and install a complete emergency lighting system as indicated on the drawings and specified herein. The system shall originate on the line side of the service-entrance main switch, through overcurrent protective equipment to each exit light fixture, and each fixture designated as being "emergency light." The switch shall be painted red. The Contractor shall be responsible for verification with local governing authorities of the proper letter and background colors of exit-light fixtures before purchase of same. The entire installation shall be in accordance with the National Electrical Code, the local electric code, and the fire-protection department having authority in the local jurisdiction.

Section 16G—Underfloor Electrical Distribution System

1. Portions of the sections of the Documents designated by the letters A, B, and C and "DIVISION ONE—GENERAL REQUIREMENTS" apply to this Division.

2. Consult Index to be certain that set of Documents and Specifications is complete. Report omissions or discrepancies to the Architect.

3. UNDERFLOOR DUCT SYSTEM

a. Furnish and install, as shown on plans, an electrical underfloor distribution system as manufactured by Walker-Parkersburg. This system shall consist of No. 4 ducts for telephone and No. 2 ducts for power service together with the necessary junction boxes, couplings, supports, adapters, and feeders to form a complete installation, made watertight with sealing compound.

b. Duct: Shall be manufactured from No. 14 gauge steel and finished with a UL-approved corrosion-resistant coating. Number 2 Duct for power service shall be $3\frac{1}{8} \times 1\frac{1}{4}$ inches with threaded 2-inch IPS inserts, spaced on two-foot centers. Number 4 Duct for telephone service shall be $6\frac{1}{2} \times 1\frac{1}{2}$ inches with two-inch IPS inserts on two-foot centers.

c. Junction boxes: Shall be Cat. No. 224-S with finish similar to duct. Cat. No. 224-S-LH Tile Holders shall be furnished with depth as required for installation in floor finish specified. All boxes shall be carefully leveled with the tops flush with the finished concrete floor.

d. Supports: All ducts shall be accurately aligned and leveled with the top of the inserts at $\frac{1}{16}$ inch to $\frac{1}{8}$ inch below the finished concrete floor. They shall be held in place during the pour by Cat. No. S-224 supports, spaced at five-foot intervals.

e. Markers: Cat. No. 415-B and No. 415-N markers shall be placed in the last insert at all dead ends, on each side of partitions, and at the first insert adjacent to the junction box.

4. SERVICE FITTINGS—UNIT PRICE

a. Provide a unit price for installing floor receptacles in the underfloor duct system. The unit price shall include: (1) One Walker-Parkersburg No. 513 ALEDPSG-G service fitting with adapters, locking nipples, or supports as required; (2) Duplex outlet as specified; and (3) the floor receptacle shall be installed at the locations as directed by the Owner complete including the wiring from the receptacle to the panel with a maximum of six receptacles per circuit.

b. Provide a unit price for installing a Walker-Parkersburg No. 518 AL-G telephone service fitting with adapters, locking nipple, or support as required.

5. SERVICE FITTINGS SHOWN ON THE DRAWINGS
Shall be as specified above and shall be part of the Contract Price.

Section 16H—Panelboard Schedules

1. Portions of the sections of the Documents designated by the letters A, B, and C and "DIVISION ONE—GENERAL REQUIREMENTS" apply to this Division.

2. Consult Index to be certain that set of Documents and Specifications is complete. Report omissions or discrepancies to the Architect.

3. Refer to drawings for panelboard schedules.

Section 16I—Emergency Standby Service

1. Portions of the sections of the Documents designated by the letters A, B, and C and DIVISION ONE—GENERAL REQUIREMENTS" apply to this Division.

2. Consult Index to be certain that set of Documents and Specifications is complete. Report omissions or discrepancies to the Architect.

3. EMERGENCY SYSTEM

a. General: The Electrical Contractor shall furnish and install a complete emergency lighting and power system, as shown on the plans, including necessary standby power equipment, control accessories, feeders, panels, branch circuits, and outlets as noted. All wiring shall be of the sizes indicated on the drawings and shall conform to the provisions of the NEC and all other local regulations covering this type of installation.

b. Generator: Furnish and install, where indicated on the drawings, an emergency standby engine-generator set rated as shown on the drawings. It shall be cushion-mounted on a heavy steel base and be free from torsional vibration. The engine shall operate satisfactorily on fuel specified. The engine shall have a 12-volt battery, a 12-volt starting motor, and a charging generator with automatic charging-rate regulator and shall also be equipped with low oil pressure, high water temperature, and automatic overspeed safety shutdown devices.

c. Instrument panel: The instrument panel on the standby unit shall contain engine oil-pressure and water-temperature indicators, battery charge-rate ammeter, start and stop buttons for manual operation of unit, manual reset circuit breaker, voltage regulator, ammeter with phase-selector switch, running-time and frequency meters.

d. Control panel: The wall-mounted control panel shall contain the necessary control equipment to automatically start the standby generator set when the line voltage drops to 70 percent of normal value, transfer the load to the generator, and re-transfer the load back to normal source when voltage is restored to 90 percent normal. The enclosed panel shall contain an electrically operated, mechanically held transfer switch, automatic engine-starting relays, and a cranking limiter to open the starting circuit after about 45 seconds.

e. The emergency standby generator shall be Onan, Caterpillar, or Katolight.

Section 16J—Signals and Communications

1. Portions of the sections of the Documents designated by the letters A, B, and C and "DIVISION ONE—GENERAL REQUIREMENTS" apply to this Division.

2. Consult the Index to be certain that set of Documents and Specifications is complete. Report omissions or discrepancies to the Architect.

3. FIRE-ALARM SYSTEM

a. Furnish and install a Fire-Alarm System as manufactured by Simplex and described in these specifications and indicated on the drawings. The system is to be wired and installed in accordance with the manufacturer's specifications and left in first-class operating condition.

b. Operation: At each stairway, exit, and other locations shown on the plans, there shall be a noncoded fire-alarm station. At each location, where shown, there shall be a bell or horn. Operating any station shall cause all sounding devices to operate continuously until the fire-alarm station has been restored to normal. It shall also be possible for those in authority to transmit a test signal from any station. The stations and sounding devices shall be connected to a control panel which shall permit a small supervisory current to pass through the entire system. A trouble bell shall also be provided and shall sound continuously in the event of failure of the main power-supply source or a ground fault of its installation wiring circuit.

c. Equipment: Install where shown a flush noncoded manual fire-alarm station. Flush stations shall mount on standard outlet boxes with single-gang cover.

(1) Install where shown on plans an underdome vibrating bell. Size and number of signals to be located so that they may be heard by all occupants of the building.

(2) Install where shown a closed-circuit fire-alarm control panel in flush wall-type steel cabinet equipped with hinged door and with lock and keys. Panel shall contain all necessary relays, meters, resistances, thermal cutout, terminals, and fuses for the control and supervision of the system. Panel shall contain number of bell and station circuits required. A trouble bell shall be provided for external connections.

(3) All interior wiring shall be in strict accordance with NFPA Codes 70 and 72 and all local electrical codes applying. Size and number of wires shall be in accordance with wiring diagram supplied by manufacturer of fire-alarm system.

(4) The Electrical Contractor shall provide and install smoke detectors, wiring and connections to magnetic door latches, and flow switches. The fire-alarm panel shall be factory-wired to accept these and any other devices specified herein or as shown on the drawings.

Appendix B

Using a Calculator

Many electrical engineers and designers in the past have rated the slide rule as one of their most important instruments, since it helped reduce long minutes of paper-and-pencil calculations to a few simple manipulations of the "slip-stick" and "runner." While the slide rule has proved indispensable since its invention in the 1850s, there is now something faster, more versatile, more compact, more accurate, and better able to solve today's engineering problems. It is the electronic pocket calculator.

It is not the purpose of this appendix to explain the fundamentals of electronic calculator operation, since this information may readily be obtained from the handbooks accompanying these devices and since practically anyone can master the modern operations of an electronic calculator in a single evening. On the other hand, this appendix will explain how to make specific basic electrical calculations. Actual directions will be given for pressing the required keys in each case.

A selected number of examples have been chosen. These are basic; by no means do they attempt to cover all possible uses of the calculator. Other examples solvable by the same processes will readily occur to the reader.

We find that the type of electronic calculator possessed by most electrical designers is an inexpensive device costing less than $100. Our discussion, therefore, will be limited to that type. However, any calculator capable of adding, subtracting, multiplying, and dividing may be employed, except for problems involving reciprocals and square roots.

Ohm's Law (dc)

1. To find resistance when current and voltage are known:

    A. Key in voltage value.
    B. Press the division key.
    C. Key in current value.
    D. Read answer displayed.

    Example: voltage = 120 V
    current = 2 A
    A. Key in voltage.   1 2 0
    B. Press the division key. ÷
    C. Key in current value.   2
    D. See display (ohms).   60.00

2. To find current when resistance and voltage are known:

    A. Key in voltage value.
    B. Press the division key.   ÷
    C. Key in resistance value.
    D. See answer (current) displayed.

3. To find voltage when current and resistance are known:

    A. Key in current value.
    B. Press the multiplication key. ☒
    C. Key in resistance value.
    D. See answer (voltage) displayed.

4. To determine value of cathode resistor for any tube:

    A. Key in desired grid-bias voltage.
    B. Press the division key. ÷
    C. Key in tube plate-current value.
    D. See answer displayed.

5. To determine value of grid resistor for a VHF power amplifier tube:

    A. Key in desired dc grid-voltage value.
    B. Press the division key. ÷
    C. Key in recommended dc grid-current value.
    D. See answer displayed.

Ohm's Law (ac)

6. In order to find current when reactance (or impedance) and voltage are known:

    A. Key in voltage value.
    B. Press the division key. ÷
    C. Key in reactance (or impedance) value.
    D. Press the ═ key and read current value.

7. To find the voltage of a circuit when current and reactance (or impedance) are known:

    A. Key in current value.
    B. Press the multiplication key. ☒
    C. Key in reactance (or impedance) value.
    D. Press ═ key and read the voltage value.

8. If it is desired to find impedance (or reactance) when voltage and current are known:

    A. Key in voltage value.
    B. Press the division key. ÷
    C. Key in current value.
    D. Press the ═ key and read impedance (or reactance) on the panel.

DC Power

9. In order to find the power in watts when current and voltage are known:

    A. Key in current value.
    B. Press the multiplication key. ☒
    C. Key in voltage value.
    D. Press ═ key and read the answer (watts) on the panel.

10. To find the power in watts when current value and resistance are known on calculators with the square key $\boxed{x^2}$:

A. Key in the current value.
B. Press the square key. $\boxed{x^2}$
C. Press the multiplication key. $\boxed{\times}$
D. Key in the resistance value.
E. Press the $\boxed{=}$ key and read the answer on the panel.

11. In order to find the power in watts when the voltage and resistance are known on calculators with the square key $\boxed{x^2}$:

A. Key in the voltage value.
B. Press the square key. $\boxed{x^2}$
C. Press division key. $\boxed{\div}$
D. Key in resistance value.
E. Press the $\boxed{=}$ key and read the answer on the panel.

For use on machines that have no square key, problems with squared factors may be solved as follows:

A. Key in current value.
B. Press multiplication key. $\boxed{\times}$
C. Key in current value again.
D. Press the multiplication key. $\boxed{\times}$
E. Key in the resistance value.
F. Press the $\boxed{=}$ key and read the answer on the panel.

AC Power

12. In order to find the power in watts in an ac circuit when current, voltage, and power factors are known:

A. Key in the current value.
B. Press multiplication key. $\boxed{\times}$
C. Key in voltage value.
D. Press multiplication key. $\boxed{\times}$
E. Key in power factor.
F. Press the $\boxed{=}$ key and read the answer on the panel.

13. To find reactive power value in watts (voltamperes, VA) when current and impedance are known on calculators with square key $\boxed{x^2}$:

A. Key in current value.
B. Press the square key. $\boxed{x^2}$
C. Press multiplication key. $\boxed{\times}$
D. Key in impedance value.
E. Press the $\boxed{=}$ key and read the answer on the panel.

14. To find reactive power value in VA when voltage and impedance are known:

A. Key in the voltage value.
B. Press the square key. $\boxed{x^2}$
C. Press the division key. $\boxed{\div}$
D. Key in impedance value.
E. Press the $\boxed{=}$ key and read the answer in watts on the panel.

AC Conversions

15. In order to determine effective (RMS) value of alternating current or voltage from peak value:

A. Key in voltage value or peak current.
B. Press the multiplication key. ⊠
C. Press the division key. ⊡
D. Key in 0.707.
E. Press the ⊟ key and read effective (RMS) value on the panel.

16. To determine the average value of alternating current or voltage from peak value:

A. Key in peak current or voltage value.
B. Press the multiplication key. ⊠
C. Key in 0.636.
D. Press the ⊟ key and read average value on the panel.

17. To determine peak value of alternating current or voltage from effective (RMS) value:

A. Key in effective (RMS) value.
B. Press the multiplication key. ⊠
C. Key in 1.414.
D. Press the ⊟ key and read peak value on the panel.

18. To determine peak value of alternating current or voltage from average value:

A. Key in average value.
B. Press the multiplication key. ⊠
C. Key in 1.57.
D. Press the ⊟ key and read peak value on the panel.

19. When the effective (RMS) value of alternating current is desired or voltage from average value:

A. Key in average value.
B. Press the multiplication key. ⊠
C. Key in 1.11.
D. Press the ⊟ key and read peak value on the panel.

20. To determine average value of alternating current or voltage from effective (RMS) value:

A. Key in effective (RMS) value.
B. Press the multiplication key. ⊠
C. Key in 0.9.
D. Press the ⊟ key and read average on the panel.

Reactance

21. If it is desired to know the reactance of a coil of negligible resistance when inductance and operating frequency are known:

A. Key in 6.28.
B. Press the multiplication key. ⊠
C. Key in frequency.
D. Press the multiplication key. ⊠
E. Key in inductance value.
F. Press the ⊟ key and read reactance on the panel.

22. To find the reactance of a coil of negligible resistance or of a capacitor when the line voltage is known as well as the current through it:

A. Key in voltage value.
B. Press the division key. ÷
C. Key in current value.
D. Press the ⊟ key and read reactance on the panel.

23. To find the reactance of a capacitor when capacitance and operating frequency are known:

A. Key in 6.28 .
B. Press the multiplication key. ⊠
C. Key in frequency.
D. Press the multiplication key. ⊠
E. Key in capacitance.
F. Press the ⊟ key and read reactance on the panel.

Unlike the slide rule, the use of the electronic calculator automatically places the decimal point in the answer. Therefore, when using the calculator, it is not necessary for users to keep before them a mental image of the proceeding arithmetic in the sense that they visualize the working of the example on paper.

It is recommended that the readers first become familiar with the particular type of electronic calculator they plan to use. Then substitute actual values in the preceding examples in order to become familiar with the procedures involved in the calculations.

Index